KETTERING COLLEGE
MEDICAL ARTS LIBRARY
DISCARDED
D0853249

RESISTANCE

# RESISTANCE

## THE HUMAN STRUGGLE AGAINST INFECTION

BY NORBERT GUALDE

*Translated by Steven Rendall*

Washington, D.C.—New York

9/07

Originally published as *Les Microbes aussi ont une histoire*
© Les Empêcheurs de penser en rond/
Le Seuil, 2003

Copyright 2006, this translation, Dana Press, all rights reserved
Published by Dana Press
New York/Washington, D.C.

The Dana Foundation
745 Fifth Avenue, Suite 900
New York, NY 10151

900 15th Street NW
Washington, DC 20005

DANA is a federally registered trademark.

ISBN-13: 978-1-932594-00-3
ISBN-10: 1-932594-00-0

Library of Congress Cataloging-in-Publication Data

Gualde, Norbert.
[Microbes aussi ont une histoire. English]
Resistance : the human struggle against infection / by Norbert Gualde ; translated by Steven Rendall.
p. ; cm.
Includes bibliographical references.
ISBN 1-932594-00-0
1. Drug resistance in microorganisms. 2. Epidemiology. 3. Emerging infectious diseases. 4. Natural immunity. 5. Human ecology. I. Title. II. Title: Human struggle against infection.
[DNLM: 1. Disease Outbreaks—history. 2. Communicable Diseases, Emerging—epidemiology. 3. Immunity, Natural. WA 11.1 G911r 2006a]
QR177.G8313 2006
616.9'041--dc22
2006019581

Cover Design by Andrea DuFlon
Text Design by G&S Type

Support for the publication of this book was provided by the French Ministry for Culture—Centre National du Livre.

www.dana.org

# CONTENTS

INTRODUCTION **vii**

CHAPTER 1
Histories of Epidemics: Relationships Between the Immunities of Populations and the Environment **1**

CHAPTER 2
The Emergence of New Diseases **29**

CHAPTER 3
The Return of Old Diseases and Persistent Diseases **69**

CHAPTER 4
The Immunity of *Homo Sapiens:* From the Individual to the Group, from Biology to Culture **99**

CHAPTER 5
Gaia or Chaos **123**

EPILOGUE
Man is the Epidemic **149**

# INTRODUCTION

*. . .the vast immeasurable Abyss*
*outrageous as a Sea, dark, wasteful, wild . . .*

MILTON, *PARADISE LOST*, VII, 211–212

Between 1950 and 1970 it was thought, in medical circles and among public health officials in industrialized countries, that infectious diseases would inevitably be wiped out. This optimism was also endemic in the Geneva offices of the World Health Organization (WHO). Thanks to advances in hygiene, vaccinations, and antibiotics, microorganisms harmful to human health would be consigned to the dusty pages of the histories of medicine and epidemics. Like the smallpox virus, they would someday exist only in reference laboratories transformed into eco-museums of infectious agents. There would be no more typhus, no more typhoid; meningitis would be cured, sanitariums closed; even the pediatric departments in hospitals would be empty because of the decrease in the number of infectious childhood diseases. There would no longer be any reason to earmark substantial research monies for studies on transmissible diseases, or to allocate new funds for hospital units where these illnesses would cease to be seen. Lacking job opportunities and funding, young specialists in infectious diseases would leave the universities to enter other fields of biological science. The science, medicine, technology, and culture of the twentieth century would be able, it was thought, to eradicate the most dangerous microorganisms. People would live without worries in an artificial, comfortable, and reassuring immunity.

Those who experienced the tragedy of tuberculosis before the use of antibiotics can understand the immense optimism of the 1960s, which was encouraged by repeated discoveries of new antibiotics (by 1965, there were twenty-five thousand of them). Penicillin, which came into use in 1944, was seen as a miracle drug, but it was not the only discovery that gave people confidence in the future. For example, by eliminating lice, DDT (dichlorodiphenyltrichloroethane) made it possible to eradicate typhus epidemics and, by killing the *Anopheles* mosquito, to hope that malaria might be eliminated as well. Vaccines also inspired great hopes. This confidence was certainly legitimate; it was based on the decline in smallpox, the decrease in cases of poliomyelitis in North America and western Europe (from 76,000 in 1955 to 1,000 in 1967), and the usually easy cures available for most bacterial infections, including tuberculosis. However, instances of bacteria resistant to antibiotics were recorded as early as the 1950s. It was already understood that users of these miraculous drugs had artificially selected microbes with genetic equipment allowing them to nullify the effects of the molecules that were supposed to destroy them.

Initially, resistance to antibiotics did not arouse serious concern, because people were confident that research being carried out by pharmaceutical laboratories would discover new substances capable of getting around these microbial eccentricities. It was, and still is, a question of finding new drugs to counter microbes seeking to resist them. This battle, which has neither victor nor vanquished, makes the treatment of infectious diseases difficult, complex, and expensive. The cat-and-mouse game in the microscopic domain has led to the emergence of microbes that adapt themselves to the situations imposed on them; they are veritable Proteuses of resistance to antibiotics. The germs in question, called resistant or multiresistant bacteria, often endanger the lives of those whom they attack. The hope that we would be able to eradicate infectious diseases is now lost. In the future, we will need systems of detection that allow us to discern the conditions in which dangers due to microorganisms are likely to emerge.

Acquired immunodeficiency syndrome (AIDS) illustrates the problem of the emergence of so-called "new" viruses, even if there are good reasons for thinking that these new viruses are to viruses what the New World is to the world. Since the appearance of AIDS, everyone sees infectious diseases as a potential danger, a real threat. Our perception of "plagues" is coming closer to that of earlier generations. Isn't it amazing that viruses were forgotten for decades in this way? It was as though the worldwide eradication of smallpox was ipso facto proof that an easy victory would be won over all other viruses. People forgot, it seems, how catastrophic a flu epidemic can be. Strangely, the epidemic of Spanish flu that occurred just after World War I received relatively little attention from the media, even though it killed more people than the Great War itself. The epidemics of smallpox, influenza, AIDS, and other diseases caused by viruses raise questions to which there are no easy answers: How did they happen? Why these viruses? Why at this time and not another?

It is clear that human activities are sometimes responsible for the emergence of so-called new diseases or for the reactivation of quiescent infectious agents. Some of these behaviors, such as wars, are episodic, whereas others lead to modifications of the natural environment, repeating themselves and becoming chronic, regular, and continuous. Deforestation is an example of an economic choice with devastating consequences for ecology and microbial dangers. It is well established that humans have brought out viruses and parasites from the depths of the forests, thus promoting the diffusion and proliferation of pathogenic agents that were previously located far from any usual human activity. We will reconsider cases of epidemics following anthropic modifications of the environment such as the one that took place in Bolivia in 1962. The hemorrhagic fever produced by the Machupo virus resulted from a program of agrarian reform that had been put in place ten years earlier. The Bolivian government allowed peasants to cut down the forest in order to plant corn. The land clearing was accompanied by a use of DDT so massive that it killed cats as well as mosquitoes. This was a real godsend for the *Calomys* mouse, which found itself in a mouse's paradise that had no predators and was full of food. The animal proliferated and through its urine spread the virus, which was harmless for it but lethal for human beings. The Argentine hemorrhagic fever caused by the Junin virus resulted from the use on the pampas of herbicides that were intended to increase corn crops but also allowed the multiplication of a plant that the *Calomys* mouse loved to eat. We can guess what happened: the mouse carrying the Junin virus proliferated and transmitted the virus to the inhabitants of the region. Disturbance of the environment is, or is thought to be, involved in the emergence of the Oropouche virus, the Marburg, Lassa, and Ebola fevers, and perhaps even in that of the human immunodeficiency virus (HIV).

An undeniable threat is posed by apparently new infectious agents and by other agents expert in resisting therapeutic measures or in getting around organisms' immunity (defense systems). Media looking for sensational news regularly put stories about AIDS on their front pages, even though worldwide, malaria kills at least as many people every year as AIDS has since it first appeared. The wealthy countries' information media must not cause us to forget that despite the search for new medicines and testing of vaccines, the malaria epidemic continues to develop dramatically, because the parasite is resistant to drugs and its vector, the *Anopheles* mosquito, is resistant to insecticides. At the end of the twentieth century, two billion people were exposed to malaria, five hundred million to one billion people were contaminated, and one million to two million died from the disease every year, most of them children. And we mustn't forget prions, the fascinating agents responsible for mad cow disease, which is now known to be a new variant of Creutzfeldt-Jacob disease that affects people younger than those who develop the "classical" form of the disease. Prions are not even living beings, only microproteins, but they are transmissible—that is, "infectious"—between different species. Prions resist everything, including heat, and they destroy the central nervous system, leading to debilitating diseases that are always fatal.

The 1960s' confidence regarding microbes is no longer appropriate, but this does not mean that the future is necessarily bleak. We should remain vigilant. What can we propose to ward off the attack of a microorganism? We belong to a species whose numbers are constantly increasing in an exponential demographic explosion particularly worrisome because of its consequences on the environment. Taking this demography into account, we must neither surrender to a Malthusian pessimism, according to which the arithmetical increase in the food supply is inadequate to provide for the human species' geometrical growth, nor indulge in an Enlightenment optimism that assumes humans will find a solution to every problem. It is recognized that an increase in the number of individuals promotes the diffusion of microbial diseases. In *The Scarlet Plague,* Jack London wrote: "The world was full of people. The census of 2010 gave eight billions for the whole world. . . . yes, eight billion people were alive on the earth when the Scarlet Death began." In writing those words, was this talented writer manifesting a visionary gift? Let's not conceive of the environment anthropocentrically. Man constitutes the environment as quality, and contributes to the equilibrium and disequilibrium of the environments of which he is a part. Like an ecological Janus, he destroys and improves his ecosystem; he is both a polluter and a victim of his own pollution.

Against the dangers of which he is simultaneously the cause and the victim, man opposes his natural powers of resistance, which are biological in nature and constitute his immunity. For human groups, the immunity of each individual participates in the immunity of the human population to which he belongs. To this must be added cultural defenses—that is, the set of concepts and technological means that allow us to compensate for the gaps in biological immunity. The immunology of populations belongs to a level of integration that associates biology with culture. In Henzel Henderson's formulation, the immunology of populations is the product of the evolution of biological DNA and "cultural DNA," with which we could connect Richard Dawkins's idea of the *meme*. The cultural role of the immunities of populations was stressed in a way by Jack London, whose hero escapes destruction in the scarlet plague epidemic thanks to his biological immunity; he bequeaths to his grandchildren, who have returned to the primitive stages of *Homo sapiens,* his own books accompanied by a method of reading them.

In a period of microbial danger, the key terms are not only biological immunity and cultural immunity but also evolution and complexity. Human biological evolution has accompanied that of microbes; there is a coevolution of *Homo sapiens* on one hand, and of microorganisms in his environment on the other. It goes without saying that this assertion is only partially valid for new germs that have not yet discovered man and that man has known about only for a short time. A single person constitutes the environment for the billions and billions of bacteria that inhabit his skin and digestive system. This coevolution proceeds from permanent conflicts between the immune system and microbes, and results from ongoing interactions (cf. Chapter 5). It has promoted the establishment and maintenance of a polymorphism of human groups, a polymorphism one of whose most elaborate forms is represented by the major histocompatibility complex—that is, the group of genes that help the immune system recognize foreign substances. Discussing the biological evolution of microorganisms, ecosystems, and human beings, Murray Gell-Mann claims that mammals' immunity is one of the most nearly perfect complex adaptive systems. In Gell-Mann's representation of the system, the biological stage precedes that of learning and thought, which in turn leads to the cultural stage in humanity's development.

In short, it is tempting to imagine the world of living beings as functioning in a harmonious and almost magical way, as is proposed by the concept of Gaia, named after the Greek goddess of Earth. Harmony can be broken by harmful microorganisms destroying the equilibrium and replacing it with a chaotic system governed by its own determinism. According to Greek mythology, Earth took the place of Chaos; Earth was not related to Chaos but they were connected to each other by supernatural events at the origin of the world.

Biological and cultural evolution will continue; what will emerge from them in the event of microbial dangers? Will we see a devastating cataclysm that will expose us to a worldwide holocaust? Will there be repeated episodes of acute attack occurring against a "background noise" of chronic infections? If our planet is a living being, it will be largely a matter of a violent attack from the inside, a kind of self-destruction or cancer against which James Lovelock's favorite organism will have to find the means to maintain its homeostasis. In his *The Ages of Gaia,* Lovelock writes that the arrival of *Homo sapiens* modified the nature of the planet. Anthropic modifications are ongoing, and we do not know where they will lead. Man can corrupt the environment, but he can also protect it. The result of this frequently observed duality will depend on measures that are often difficult to decide upon and even more difficult to put into application.

The relationship between humans and microbes is a complex, dynamic phenomenon; it is not linear and often corresponds to a chaotic and sometimes wholly unpredictable determinism. We may seek in vain to harmonize Gaia's laws with those of chaos; there will always be an element of unpredictability in this complicated relationship.

CHAPTER 1

# HISTORIES OF EPIDEMICS: *Relationships Between the Immunities of Populations and the Environment*

*"Damn it! And whooping cough, you can get that just by looking at somebody! It's a kind of flying microbe, hundreds of millions of times smaller than a mosquito! Even if a doctor showed it to you and said: "there it is," well, you could look all you wanted and you wouldn't see it. It's a monster with terrible fangs . . . And as soon as it sees a little child, the bug jumps on him and tries to eat his throat, causing him miseries without end!"*

MARCEL PAGNOL

## SMALLPOX AND THE IMMUNITIES OF POPULATIONS: INOCULATION, VACCINATION, AND THE WORLD HEALTH ORGANIZATION'S VICTORY

### *The heroes of novels, kings, "savages" . . .*

Readers of Zola's *Nana* can never forget the description of the terrible death that carries off a romantic heroine of pleasure and voluptuousness; we always remember how "*Venus decomposed.*"[1] We see why, in order to illustrate one aspect of the social determinism in his work, Zola used a disease and its virus whose origin is lost in the depths of time. Smallpox has always inspired the worst horror in our species, and it's true that the epidemics smallpox has caused have left in our collective consciousness hideous, macabre memories comparable to the fears aroused by plague and cholera. For this reason, smallpox might rightly replace plague in representations of the third horseman of the Apocalypse.

In May 1774, King Louis XV met an end similar to that of Zola's heroine. He was abandoned by everyone, even Madame du Barry, his current mistress. In order to abridge a visit to the sovereign that seemed very dangerous, the archbishop of Paris, Monseigneur de Beaumont, claimed to be suffering from renal colic. Smallpox did not limit itself to striking down princes and heroines of novels. It was an illness strictly limited to our species, and it affected people of all social classes. Smallpox has had calamitous epidemic consequences; it has played a major role in the disappearance of certain human groups and in the collapse of civilizations.

The smallpox virus,[2] which belongs to the Pox group, affects no living species except the human species, in which it produces a highly contagious and epidemic illness. The human nature of the virus's preferred terrain explains how it was possible to do away with the disease by vaccinating the great majority of the world's population. The last case of the minor form of the disease, still called alastrim, was described in October 1977 in Somalia, and the last case of the major form of the disease was that of the child Rahima Banu in October 1975 in Bangladesh.

The incontestable success constituted by the eradication of an infectious disease responsible (according to the WHO) for about two million deaths a year in the 1960s must not cause us to forget what this calamity has meant in the course of humanity's history.

People had imagined, in a very speculative way, that *Homo neanderthalis* might have disappeared as a result of a smallpox epidemic. Alongside this hypothesis, which it is impossible to confirm or disconfirm scientifically, there are other historically verified examples attesting to the terrible demographic effects of the disease during the history of humanity. For eighteen hundred years, smallpox regularly killed two hundred thousand to six hundred thousand people per year in Europe, and that means that over the centuries it has killed more people of all ages, social classes, and ethnic groups than any other infectious disease.

No one denies that smallpox is a disease whose pressure on humanity was such that it had profound historical consequences. In November 1518, Cortés left Cuba for the Yucatán, reaching the city of Tenochtitlán (now Mexico City) on November 8, 1519. During the three subsequent years, half the population of the city is supposed to have been carried off by smallpox, a demographic collapse announcing the beginning of what has been called an "unpremeditated genocide." The Spanish colonization of South America, the *Conquista*, was incontestably made easier by smallpox. After the Spanish arrived, the Amerindians suffered smallpox epidemics for decades, the virus finding in Amerindian America—which was to become Latin America—a population in which the disease was previously unknown. Completely uninfected by smallpox, the Amerindian population had not had to adapt itself to the selective pressure exercised by the disease, whereas Spaniards had already had to "pass the bar" of the danger of smallpox in order to live into adulthood. Smallpox, a circumstantial ally of the Spanish, who had already grappled with the disease, provoked a demographic crisis so severe that by 1620 Mexico had only 2 million inhabitants, whereas the population before the Spanish arrived has been estimated at 20 million to 30 million. To be sure, smallpox does not explain everything; other infectious diseases such as measles also played a role in the disaster. Contemporary witnesses mention the particular effects on the Aztecs of the white man's military technology combined with his diseases. The whole experience generated among the Mexicans the impression that they were confronted by a calamity of divine origin, and the *Conquista* turned the Aztecs into a people resigned to their fate.

Epidemics were to strike the continent again in 1531, 1541, 1564, 1576, and 1595, directly or indirectly causing the death of 56 million people, affecting the Aztecs, the Mayas, and the Incas.

Similarly, the smallpox epidemic that struck Easter Island late in 1863 had killed half the island's inhabitants by the end of the following year. In fact, whether smallpox epidemics were European, Asian, African, or American, a high rate of mortality was regularly observed; as many as 50 percent of those who fell sick died. In such a deadly context, the case of the Amerindians was special, because they proved to be exceptionally sensitive to viruses coming from Europe. This fact brings us to the notion of the immunities of populations.[3]

In order to deal with the great killer (40 to 50 percent of those infected by the disease died, the survivors bearing forever the indelible scars of the "pock-marked"), people came up with countless therapeutic approaches as ineffective as they were bizarre, strange, or absurd. Sometimes the more-or-less improvised efforts of therapists were more rational, particularly when they used a technique of prevention based on inoculation. The principle was simple: since the victims who survived enjoyed permanent protection against any new attack by the virus, it sufficed to inoculate healthy people with liquid taken from the pustules in the hope that this would result only in a minor form of smallpox while at the same time providing permanent immunity to the disease.

### *Inoculation, the first true artificial immunization, preceded true vaccination*

The technique of inoculation that came from the East and the Caspian Sea area consisted of administering the virus taken from lesions that were already healing, using a needle, or, as was done by the Chinese, inhaling the contents of a swab impregnated with liquid from pustules, which produced an inoculation through the nasal mucous membranes. In 1717, Lady Mary Wortley Montagu, the wife of the English ambassador to Turkey, had her only son inoculated, with the result that the method became widely known in western Europe. This early method of inoculation consisted of deliberately administering the smallpox virus, a practice whose undesirable and potentially very serious effects made it a dangerous form of protection. Thanks to Edward Jenner, inoculation was replaced by vaccination. On May 14, 1796, Jenner carried out the first vaccination by administering to an eight-year-old boy a vaccine taken from a dairymaid. The boy became resistant to the inoculation, that is, to the smallpox virus. Resistance to smallpox (and therefore to the inoculation) was known among those who tended cows, who, when they came into contact with cowpox (the vaccine) got a benign illness that protected them from smallpox. In short, inoculation consisted of using the smallpox virus to immunize the subject, whereas vaccination drew on the virus of the vaccine (a relative of the smallpox virus), thereby providing immunity (vaccination) without the illness (smallpox). Let us put it this way: vaccination involves injecting the virus of the vaccine, which produces a benign, localized illness that results in an immunity to smallpox in the person vaccinated.

### *The World Health Organization's victory*

In the 1960s, vaccination was practiced worldwide in the framework of a program whose aim was to eradicate smallpox and which was administered by the WHO. The program was carried out between 1967 and 1977. The goal of this war on smallpox was to vaccinate 80 percent of the world's population, the percentage of immunization considered necessary to eradicate the disease. The result was remarkable. On October 26, 1979, in Nairobi, the WHO's general director declared that smallpox had been eradicated from the planet and proposed to call that day "Smallpox zero day." On May 8, 1980, it was announced at the WHO's thirty-third general assembly that smallpox had disappeared. This amazing victory came only 177 years after Jenner's discovery, but 3,135 years after the case of smallpox that killed the pharaoh Ramses V. It is interesting to note that the campaign to eradicate smallpox cost American taxpayers 300 million dollars, whereas sending a man to the moon cost them 24 billion dollars.

The preceding does not mean that we are safe from any kind of smallpox or any smallpox-like disease. The disease has disappeared, but not the virus; it could still be present in the remains of humans buried in the permafrost or in mummified bodies. By agreement with the WHO, the virus is officially preserved in two world laboratories. Initially, depositing the virus in these reference laboratories (the Vektor Center at Koltsovo, near Novosibirsk, Russia, and the U.S. Centers for Disease Control and Prevention in Atlanta, Georgia) was supposed to make it possible to preserve the virus for the purposes of scientific research. The stored viruses were supposed to be eventually destroyed, but this destruction has not yet been carried out. Opponents of destroying the viruses note that there might be "forgotten" viruses in laboratories or elsewhere (in the permafrost, human remains, etc.) and that this explains why the reference stocks must be preserved in the event that unfortunate circumstances arise. It is always useful to have an intimate knowledge of dangerous viruses. For example, thanks to the patient work done by molecular biologists Jeffrey Taubenberger and Ann Reid on autopsy samples taken from the bodies of flu victims preserved in the Alaskan permafrost, it has been possible to discover the genes and the origin of the virus that caused the 1918–1919 flu pandemic (cf. J. K. Taubenberger, Ann Reid, and T. G. Fanning, "Capturing a Killer Flu Virus," *Scientific American,* January 2005, 48–57). If the destruction is carried out, it will be the first time that a biological species has been deliberately eliminated (unless the virus is secretly preserved in military laboratories).[4] In his book *Ces virus qui détruisent les hommes,* Claude Chastel stresses the dangers involved in preserving the virus.[5]

We cannot discuss the potential danger of smallpox without mentioning its close relations in the family of pox viruses, which include whitepox and monkeypox. The latter illness, producing in man a smallpox-like eruption, can sometimes be serious or even fatal. A recent emergence of this disease in Africa has alarmed WHO authorities.[6] It is pertinent to ask whether monkeypox could be the cause of a new form of smallpox.

In summary, since the celebrated "Smallpox zero day," smallpox, a human disease with an exclusively human reservoir, has been eliminated, but is nonetheless probable that the virus is still present here and there in our natural environment. Considering smallpox's resistance, reactivation by human contamination remains possible.

## PLAGUE, OR THE RELATIONSHIPS BETWEEN A BACILLUS, FLEAS, RATS, AND MEN

In the Prado museum in Madrid, there is a picture by Breughel the Elder representing plague, the third horseman of the Apocalypse, in the shape of a skeleton on a horse pulling a cart full of plague victims. Plague, and all it symbolizes, was also used by great writers, poets, philosophers, and moralists, including the likes of Boccacio,[7] La Fontaine, and Camus. Apart from Defoe for the plague in London and Pagnol for the plague in Marseilles, these authors set out to describe not the disease itself but rather the problems encountered by their heroes in the very special context of an epidemic of the pestilence. A writer in search of a dramatic situation with apocalyptic overtones could hardly make a better choice. The great epidemics of the Black Death in fourteenth-century Europe, whose effects have been compared with those of a nuclear war, must indeed have seemed apocalyptic to those who experienced them. Fernand Braudel's descriptions report more soberly what the epidemic was probably like in reality.[8] Plague is a disease that has been known for a very long time, particularly in Europe. In some cases, as in that of the plague in Athens, there is debate as to the nature of the disease, but in general it is agreed that in Europe smallpox was regularly present. It experienced another upsurge in the fifteenth century and then disappeared, mysteriously, in the eighteenth. In France, the smallpox virus was not the only regicidal microorganism; for example, St. Louis (King Louis IX of France) is often said to have died of plague in 1270, during the siege of Carthage. Although historians no longer think he died of plague—he probably died of dysentery instead—it is unlikely that this will change popular belief regarding the cause of death of the "Holy King."

Plague is of bacterial origin. The bacillus in question was discovered in the former French Indochina by a Pasteur Institute researcher named Alexandre Yersin, whence its name, *Yersinia pestis*. It also strikes rodents, and this is important for understanding the rest. Among human beings, the symptoms differ, depending on whether the plague is pneumonic, ganglionic (bubonic), or septicemic. Contamination may take place between humans, in the case of pneumonic plague, or occur through the intermediary of fleas living as parasites on infected rats. Usually, the bacterium that infects human beings has been previously carried by a host animal, the rat (*Rattus rattus* or *Rattus norvegicus*), and is then inoculated into the future human victim by a flea (*Xenopsylla cheopsis*). When the flea passes from the animal to the human, the degree of the inoculation will depend on how extensive the infestation is and the number of bites it leads to. The gravity of the disease is also connected with the virulence of the germ when confronted by the victim's defense systems. Plague epidemics have had such enormous consequences that their advent is mentioned in the Bible[9] and by historians,[10] writers on socioeconomics, and, of course, writers on medicine and epidemiology.

### *The Black Death in Gascony and elsewhere*

The fourteenth-century attack, which is in every respect exemplary, may be taken as a model of the epidemic. The disease that struck Europeans in 1348 is supposed to have first appeared in China, or perhaps in Mongolia. Biraben situates the beginning of the catastrophe in 1328, in the area around Lake Balkhash in central Asia. Near Lake Issyk-Kul, in Kyrgyzstan, cemeteries show that in 1338 and 1339 there was an abnormally high number of burials, and on some gravestones plague is mentioned as the cause of death. The points on the disease's progress through Europe are well known. Rats, fleas, and the bacterium were transported by trading ships, so that Genoa and Marseilles were attacked first, in 1347;[11] Montpellier and Narbonne in February 1348; Toulouse in April 1348, Agen in May, and Bordeaux in August of the same year. The epidemic in Bordeaux was rapidly impressed on people's minds by the death of the archbishop on August 9, 1348. The bacillus then moved through the Aquitanian peninsula, following the route of the Garonne River along a trading axis linking the Mediterranean with the Atlantic, and a favorite alternative to the dangers associated with passing through the Straits of Gibraltar—travelers preferred to take the land route through France rather than face the Barbary pirates. Thus Gascony was an important zone of commercial transit, Bergerac and Périgueux were famous markets; and there were also La Rochelle, Bordeaux, and Bayonne, Atlantic coast ports from which one could reach England, which was still the dominant nation.

### *The causes of the emergence and virulence of plague*

In the preceding, we have seen how plague traveled from Asia to the Atlantic. But what caused the emergence of plague at the disease's initial site? First, we may mention the arrival in the area concerned—that is, the environs of Lake Balkhash—of a particularly virulent germ. The appearance of a virulent form of *Yersinia pestis* might have been produced by an evolution, or rather a coevolution, of the pathogen and its usual hosts, rats and men. The British geneticist J. B. S. Haldane was the first to point out that infectious diseases participate in evolutionary processes. This is well illustrated by the Australian myxoma virus in the 1950s, when the continent's wild rabbit and the virus, which came from Brazil, reached a kind of equilibrium between attacker and attacked (at the price of a massive die-off of the little mammals and an adaptation of the virus to the resistance of those that survived). In this case, after the epidemic the rabbit was not the same as its ancestors: it was more resistant to a virus that also changed.

The explosion of the fourteenth-century plague was perhaps the result of a brutal increase in the virulence of *Yersinia pestis*. The bacterium, which had previously been subjected to the immunologic pressure of its hosts, may have benefited from a mutation in its genome that made it more aggressive in escaping the constraints presented by the combined defense systems of humans and rats. It has been shown how a mutation of a gene in the plague bacillus made it virulent, but this does not mean, of course, that the epidemic of the fourteenth century was simply the result of one of the bacterial mutations selected by the immunologic pressure of the hosts, "forcing" the plague bacillus, so to speak, to become more virulent in order to survive. Even if the emergence of a more pathogenic bacterium would have been possible, though not demonstrated, the seriousness of the epidemic itself could not be explained solely by the nature of the germ. There are circumstances of various kinds that have surely favored the extraordinary explosion of so lethal a pathogen.

When the plague arrived, France was undergoing an exceptional period of political, economic, and demographic crises. To all this, the great Black Death epidemic added a feeling of fear, panic, impotence, and insecurity. The immense majority of people saw the epidemic as a punishment inflicted by the divine. The phenomenon was viewed in an apocalyptic and almost eschatological perspective. Death came to figure in the most explicit form in numerous sacred and profane works of art. Plague arrived in Marseilles shortly after the fall of Calais (August 4, 1347), whose defenders had put up a heroic, year-long resistance. The plague struck while France, which was in the middle of the Hundred Years' War and greatly weakened, was already undergoing a demographic decline. In Aquitaine, which has particular interest for us as a model, the thirteenth century had seen a population increase connected with a higher fertility rate and a decrease in neonatal mortality. However, let us not lose sight of the fact that at that time life expectancy was thirty years of age, and that one out of every two newborns died. Cities like Bordeaux (forty thousand inhabitants) and Périgueux (ten thousand inhabitants) were overpopulated because of the exodus from the countryside toward urban centers, because the rural areas were being ravaged by brigands and armed gangs of various origins and were not very secure. In Bordeaux, for example, living conditions were terrible so far as hygiene and health were concerned. The inhabitants were piled on top of one another in tenements; they almost never bathed and changed their clothes only once a month, or twice a month at best. This lamentable overcrowding had increased, while no measures had been taken to dispose of refuse. The streets, with their central gutter drains, served as sewers, and people put all their refuse there, from wastewater to butcher's waste, including excrement and blood. The smell was overpowering. Streams such as the Peugue and the Devèze, transformed into sewers, contaminated the drinking water. Pigs and dogs were the age's only sanitation workers.

These deplorable conditions in a country exhausted by war could only favor the plague disaster. All the conditions necessary for fleas and rats (*Rattus rattus*) to proliferate were present. In addition, the climate was at that time particularly warm and humid. In short, all the required ingredients for the arrival of the disease maximized their deleterious effects. It only remained for the bacillus itself to be brought into the Aquitanian peninsula by commercial transactions and by voluntary or involuntary human migrations (pilgrimages, troops on the march, refugees).

### *Four centuries of an undesirable presence*

The plague that attacked France at the beginning of the fourteenth century did not truly leave the country (and Europe) until the eighteenth century. To be sure, the successive epidemics each came to an end. The end of the Black Death was favored first by a considerable decline in population. Bordeaux is supposed to have had thirty thousand dead, while Périgueux and Bergerac are supposed to have lost 25 to 40 percent of their populations. In addition to this mortality there was also the flight of terrified inhabitants.[12] The panic explains why in this troubled age superstitious and frightened people often accused human groups on the margins of society of being responsible for the disaster. In France, lepers and Jews were regular victims of blind reprisals.[13]

The decisions made in an attempt to save people were not all completely absurd: for instance, pertinent hygienic measures were taken that consisted in burning all refuse, improving the streets, and organizing burials.[14] During the year 1348, a large part of the port of Bordeaux (la Rousselle) was burned, and so was the Saint-Jean Bridge and the houses in Poitevin Street. The treatments that sought to cure the illness were ineffective, of course, but at least people tried to organize themselves, while creating genuine *cordons sanitaires,* represented in particular by the guardians of the city gates.

Although they are only approximate, demographic figures suggest that during the fourteenth century immunity gradually grew within the European population and that it ultimately added a biological element to the improvement in survival rates. Statistics show that with the passing of time, the slaughter diminished. Between 1348 and 1382, the number of sick declined and the rate of recoveries increased.

We can imagine that those who recovered belonged to a group of good immunologic responders selected by the disease, and that non-responders did not survive; on the other hand, the plague bacillus may also have been losing its virulence. These two hypotheses are not mutually exclusive. Statistics on the epidemic in Venice in 1630 support the hypothesis that the immunities of populations play a role in epidemics. Immunity to plague was established over the centuries by selecting the responding individuals, an evolution that was probably accompanied by a decline in the germ's virulence. Does the coevolution of the germ and the immunity of European populations explain the almost total disappearance of major plague epidemics in the eighteenth century? The answer is not easy, because in our analysis we must not forget the role of another protagonist, the rat. The European rat that was usually infected by the plague bacillus belongs to a species susceptible to the disease: the black rat (*Rattus rattus*). In 1727, for unknown reasons (an earthquake? overpopulation?) the species *Rattus norvegicus,* also called the gray rat or sewer rat, which is resistant to plague, crossed the Volga by the millions and replaced the black rat. On this occasion, the puzzled and worried inhabitants of Astrakhan in southwestern Russia observed millions of swimming rats whose "fur threw on the river an immense gray mantle undulating with the rhythm of the waves." Did the somewhat magical arrival of this rat from beyond the Volga make a decisive contribution to the disappearance of plague? No definitive answer to this question can be given.

### *After the epidemic*

The consequences of the plague epidemics, and in particular of the one known as the Black Death, were many and major; some of them are still with us. The depth of the demographic collapse, which is said to have reduced the population of Europe by a third to a quarter, is difficult to determine. It is likely that mortality rates differed from one region to another, but there is no doubt that they were high everywhere. In some places earlier population densities were not reached again for 130 to 250 years. In addition, certain professions—priests, physicians, and notaries, for instance—were affected more than others, further destabilizing the social structure.[15] Among those most exposed were the men whose job it was to bury the dead; they were often given the nickname "crows" (*corbeaux*).

The economic impact, both direct and indirect, was fundamental. Commerce collapsed; there were no more markets, no more trade. In 1350, the Bordeaux region exported only one-third as much wine as it had earlier exported. Within families of property-owners, fortunes and real estate fell to the survivors, giving rise to new economic concentrations (monopolies, as we would now call them).[16] The disappearance of part of the population was accompanied by a loss of competence and a rarefaction of skilled labor. For example, hardly any agricultural workers survived. Those who did demanded very high salaries for working property owners' lands. This situation led English property owners to abandon farming for stock raising. Salaries thus rose for some people, but overall poverty increased.

The political consequences were just as important, and had an effect on the Hundred Years' War. In 1348, for example, the bishop of Bayonne, Pierre de Saint-Johan, negotiated an alliance between England and Castille that was to be strengthened by a planned marriage between the Spanish crown prince and Jeanne de la Tour, the king of England's daughter. The marriage was supposed to ensure that the English-Castilian fleet would have maritime supremacy along the French coast. But Jeanne de la Tour died of plague on August 3, 1348, and the project failed.

The psychological consequences ranged from terror and a sense of impotence to desperate violence and the strangest beliefs. The effect on spiritual life was all the greater because religious institutions had been decimated. People called into question the foundations of Catholic religion and prepared the way for the Reformation or sank into the most intolerant mysticism. Among the saints providing protection against the plague was St. Gregory, who, as the newly elected pope in 590, organized a procession that was said to have produced a heavenly sign of the disease ending. On the top of the well-known Castel Sant'Angelo in Rome there is a sculpture of an angel that represents the vision St. Gregory is supposed to have had during this famous procession to end the plague. We may also mention in this connection St. Sebastian—the arrows piercing his body were seen as representing divine arrows in the form of the plague—and St. Roch, who was usually represented with a bubo (a swollen lymph node) on his groin.

So far as culture is concerned, the plague was to stimulate the creativity of painters, sculptors, and writers. In France, Latin was gradually replaced by the vernacular as the vehicle of culture.

The preceding discussion has sought to summarize the history, circumstances, and consequences of a cataclysm. If it is true, as Biraben says, that "diseases have a history," it remains to be seen whether the events observed during the epidemic of the fourteenth century are in accord with those of our own time. The answer to this question is yes, since epidemics are essentially, fundamentally human. In the case of the Black Death, there are biological and cultural constants that constitute its modernity—a modernity that is all the more notable in that a strain of *Y. pestis* resistant to antibiotics has recently been discovered. What will happen if rodents begin to carry resistant bacilli? Dangerous microbes are not a recent phenomenon. Bacteria were the first inhabitants of the earth, so *Homo sapiens* and other members of the genus *Homo* have always had to adapt to threats posed by microorganisms.

### CHOLERA, THE DISEASE OF CONTAMINATED WATER, SOCIAL AND ECOLOGICAL DISORDER, AND POVERTY. SUSCEPTIBILITY TO THE DISEASE.

Cholera, known as the "Blue Death" because it gave its victims' skin a distinctive color, is another scourge that has persisted in the modern age. In the case of cholera, we should speak of pandemics rather than epidemics. Emergences of the disease are characterized by their almost constant association with social and/or ecological disorder. Moreover, we now know that cholera is carried by human individuals with genetic traits that are clearly associated with susceptibility to the disease. However, genetic predisposition should not prevent us from seeing that in our own time cholera is an illness of the poor and is associated with environmental damage.

In 1832, in France and especially in Paris, cholera struck the poorest inhabitants. This was a French case of an infectious disease whose history has been marked, for a century and half, by endemic persistence in the Ganges delta, and since the beginning of the nineteenth century, by pandemics associated with trade and human migrations.

Cholera is an infectious form of diarrhea, like typhoid, dysentery, and viral diarrheas. The bacterium that causes it is *Vibrio cholerae,*[17] usually transmitted through contaminated water. The microbe takes up residence in the bowel and secretes a toxin that adheres to the cells of the intestinal wall and causes a tremendous outflow of fluid from the body into the intestinal tract. The result is diarrhea that can cause the body to lose as much as a liter of water per hour and lead to lethal dehydration. The vibrio is found in stools, and this makes diagnosis of the disease possible. The treatment of cholera involves rehydration and antibiotics, which enable us to shorten a recovery that would be longer—if it were not cut short by metabolic disturbances. In fact, the diarrhea produced by cholera brings about a serious hydroelectrolytic imbalance that is often lethal. Cholera goes hand in hand with hygienic deficiencies and overpopulation; it is no accident that in Europe the disease has usually struck the poorest neighborhoods in cities and that it usually affects countries that lack systems of sanitation. In France, cholera was a social and vocational illness affecting the poor and especially workers laboring in a polluted area. Cholera existed before the pandemics of the nineteenth century; it was present in India when Vasco da Gama arrived there in 1498. One of his officers, Gaspar Conca, described how the disease struck certain members of the crew. A Sanskrit text dating from 500 B.C. (*Shushruta Samshita*) describes the disease, as do two-thousand-year-old Greek texts. Nonetheless, experts believe true pandemics of cholera are more recent.

### *Pandemics of the Blue Death*

The first of these modern pandemics occurred between 1817 and 1823, and its emergence was associated with two wars: the war in Oman and the war between Persia and Turkey. The second (1829–1851) began in Russia and traveled to North America, where it reached New York in 1832, later spreading to Philadelphia and New Orleans. The disease reappeared in London during the summer of 1849.[18] The third lasted from 1852 to 1859, the fourth from 1863 to 1879, the fifth from 1881 to 1896, and the sixth from 1899 to 1932. The seventh pandemic began in 1961 and is still going on; the agent in this case is *Vibrio cholerae* 1 El Tor. In July 1991, there were 45,159 known cases in ten African countries, resulting in 3,488 deaths. The vibrio 01 and the related 0139 are pathogens that can survive very well in waters of variable salinity as well as in seawater. In Bangladesh, these pathogens cohabit with plankton in brackish waters and estuaries, where they are associated with zooplankton and shellfish.

Cholera is notably affiliated with the sea. The great cholera pandemics have occurred along the world's seacoasts. The first six pandemics began in Bangladesh (Hindustan); they involved the 01 biotype. The seventh pandemic differed from the preceding ones in that it involved the El Tor biotype and began in Indonesia, spreading to Bangladesh in 1963, to India in 1964, to the ex-USSR in 1965, and to Africa in 1970. The seventh pandemic was also the only one to affect South America, beginning in Peru in Chancay, sixty kilometers north of Lima, and then spreading along the coasts of Peru, Chile, and Ecuador. The emergence of cholera in Latin America was initially attributed to a ship whose ballast, which had been loaded with contaminated seawater off the coast of Bangladesh, was emptied off Lima. Now there is general agreement that the *El Niño* current was also involved.[19] By raising water temperature and causing rains that carried into the sea soil nutrients that led to a buildup of phosphates in the water, *El Niño* produced large numbers of plankton. It has been shown that a single minute crustacean among the phytoplankton can convey up to 10,000 vibrio organisms, or ten times the infectious dose.[20] The phytoplankton that precede the zooplankton can be monitored (by satellite, for example; cf. Chapter 5) and epidemics thus prevented.

### *Biological predisposition*

A study of statistics from the nineteenth century suggests that during epidemics in Europe certain human groups were more vulnerable than others, for various social[21] or biological reasons. During the 1854 epidemic, most of the deaths were among children and, particularly, the elderly, but excess mortality rates also especially affected men between the ages of 30 and 50 and women older than 60. A study carried out on the parish records of Coutances and Dannemois, localities in the department of Seine-et-Marne, concerning the 1832 epidemic brings out genetic factors' role in mortality rates. Since cholera remains endemic in Bangladesh, or epidemic elsewhere (India, the Philippines, Peru, etc.), it is possible to carry out studies of the correlations between genetic markers and susceptibility to the disease. Not all these studies indicate a biological predisposition, but there are good reasons for thinking that in the Philippines, as also in Asia, individuals belonging to the O blood type are more susceptible to the disease.[22] A study carried out in India confirms this result. Nothing explains the association between the O group and cholera, but it has been suggested that cholera may have contributed to the low percentage of people in the O group in Asia today.

## INFLUENZA, A PAST, PRESENT, AND FUTURE DANGER

Although influenza can be a lethal disease, the general public usually considers it an ordinary illness, often confusing it with minor seasonal infections. It is rather remarkable how much this pathology is perceived as banal, even though it has caused millions of deaths. At the end of the twentieth century, plague, smallpox, and cholera remained the archetypes of disastrous infectious diseases; flu had become almost a forgotten danger. The first years of the twenty-first century have changed our perception of epidemics. The century began with the emergence of a previously unknown virus, causing Severe Acute Respiratory Syndrome (SARS) that threatened to produce a real pandemic; then came the menace of a new avian flu, reminding us of the permanent danger posed by influenza.

### *Spanish flu and the others . . .*

On March 4, 1918, in an Army camp in Furston, Kansas, doctors saw the first cases of an influenza epidemic that was later known as the Spanish flu. The disease was carried to Europe by American soldiers going to fight in France, and as early as the end of August 1918, an outbreak of flu was noted in the French port of Brest. The name "Spanish flu" probably comes from the fact that the first deaths in the civilian population were reported in Spain, while they were not mentioned elsewhere. Probably because of the historical context in which it occurred, this terrible pandemic was not to leave behind it a dramatic memory commensurate with its severity. It is thought that the Spanish flu epidemic killed between 20 million and 40 million people worldwide—that is, many more than the First World War itself.

Histories of this epidemic cannot fail to impress us. Patients who felt their first symptoms in the morning died by the end of the day; whole families were carried off within twenty-four to forty-eight hours.[23] There were even reports of women in New York who felt the first symptoms when they got on the subway and died before arriving at their destinations. In any case, by autumn of 1919, twenty-thousand New Yorkers had died in the epidemic. In Europe, the disease came in three waves, the last of which occurred in the spring of 1919. Mortality was initially high among children and the elderly, but the second wave also affected young adults, so that the initial, U-shaped mortality curve took on the appearance of a W. The infection progressed very rapidly within populations. For example, on August 27, 1918 a freighter arrived in Freetown, the capital of Sierra Leone; by August 29, 500 of the 600 dockworkers were ill; and by the end of September, 3 percent of the population of Sierra Leone had been killed by the virus. The pandemic did not spare any country, carrying off 5 percent of Ghanaians, a whole Inuit village in Alaska, 20 percent of the population of Samoa, and several million Europeans (10 to 15 percent of the population).

Flu, a seasonal disorder, is a pulmonary disease caused by influenza viruses transmitted by infected individuals through droplets of saliva. The influenza viruses are of three types, A, B, and C, and are divided into sub-types designated as H (from hemagglutinin) and N (from neuramindase). The latter designations refer to two molecules that cover the viruses and serve as "identity cards" for each sub-type (for example, the sub-type $H_1N_1$ is different from $H_2N_8$). Localized flu epidemics are annual; pandemics or major epidemics appear every ten to fifteen years.[24] The annual epidemics, which usually occur during the winter, must be taken into account because, as Dolin says, "if the great pandemics bring out dramatically the consequence of influenza, the illnesses that occur between the pandemics are responsible for an even larger number of deaths and incidence of disease over a much longer period." The alternation between local epidemics, which are generally fairly minor, and pandemics that come every ten years or so, arises from the ongoing relation between the flu virus and the immunities of populations. Human groups infected by a virus constitute, in the long run, homogeneous populations of resistant individuals. Faced with these effective defenses, only flu viruses that have an identity card that allows them to escape the vigilance of immunity can lead to new infections. Thus, in regularly occurring episodes, "new" flu viruses emerge that are capable, at least for a time, of evading the immunities of populations. The flu virus's ability to adapt to its environment is all the greater because the human virus can exchange one or more parts of its genetic material or genome with viruses of other species (pigs, ducks, etc.). As a result of this internal remodeling, the virus often acquires an increased aggressivity and pathogenic character, a renewed and exacerbated virulence. Then humans that lack for the time being an ad hoc immunity fall victim to a new flu, to which, depending on the point on the globe where it appears, is given a more or less exotic patronymic.

### *The genetic mosaic of the virus*

Phylogenetic analysis of the nucleoid sequences of the RNA of the eight A viruses has shown that all the nucleic acids that produce flu in mammals come from a pool of avian viruses. The viruses affecting mammals evolve rapidly, but those affecting birds do not. There are periodic exchanges of genes or viruses in their totality between species, and these can result in human or animal pandemics. Swine can serve as hosts in which genetic materials from different viruses are exchanged and rearranged.[25]

The flu virus's variations constitute a conundrum for researchers seeking to create vaccines against the disease. Vaccination against smallpox was based on a single type of vaccine, whereas for flu it is necessary to produce every year a new vaccine to deal with what would otherwise be a future epidemic. It is clear that the virus can defeat the experts' epidemiological predictions, even if the vaccine produced for a given year includes the flu virus that is most common at the moment. In short, preventing flu comes down to knowing how to prepare the "right" vaccine, and in this roulette game epidemiologists usually have the winning system only because they have carefully monitored the virus's development.

Within a vaccinated group, the virus confronts the immunity of populations obtained by vaccination. Hence it is important to have chosen the right vaccine to defeat the microbe, because an epidemic can strike an average-sized country in six to eight weeks. A pandemic takes only a few months to establish itself.

## YELLOW FEVER

Yellow fever is a short-lived infectious disease caused by a flavivirus. It is a hemorrhagic viral disease, historically important but also instructive, since a certain number of new viral diseases are also hemorrhagic viral fevers.

In 1886 a Cuban physician, Carlos Finlay, published a paper arguing that the *Aedes aegypti* mosquito was the vector for yellow fever. Finlay persuaded Walter Reed, head of the U.S. Army Yellow Fever Board that came to Cuba during the Spanish-American War, to investigate the mosquito's role. While Reed was in Cuba, yellow fever broke out in the United States garrison in Havana, and Reed, James Carroll, Aristides Agramonte, and Jesse Lazear were appointed to a commission to look into it. On August 27, 1900, an infected mosquito was allowed to feed on Carroll, who developed yellow fever; Lazear was also bitten, developed yellow fever, and died. Controlled experiments were then begun using volunteers, and Reed was able to prove that the infectious agent was transmitted by the *Aedes aegypti* mosquito, just as Finlay had claimed.[26] In 1901 U.S. Army engineers began a mosquito eradication program and within ninety days Havana was free of yellow fever. However, the disease remains very prevalent in Africa, particularly in West Africa, and it is also present in the Americas, where it is endemic in Central and South America.

The disease exists in three forms. The first is urban, the disease being transmitted from one person to another by the *Aedes aegypti* mosquito.

The second, sylvan form is common in Central and South America, where it is transmitted by different mosquitoes (of the genus *Haemogogus* or *Sabethes*). The third is the jungle form. In East Africa, where apes constitute reservoirs for the disease, there are other mosquitoes that attack primates and can spread the virus from animals to humans (*Aedes africanus* and *Aedes simpsoni*). Currently it is thought that apes are the natural reservoir for yellow fever and that mosquitoes are the vector.

As early as the seventeenth century, yellow fever was already a feared danger in the Americas. The first epidemic in the New World dates from 1648 and followed the arrival, aboard a ship, of the mosquito *Aedes aegypti* that was carrying the virus. It is likely that the slave trade facilitated the intercontinental transfer of the disease from Africa, where it was and remains endemic. In Senegal and Nigeria, popular songs celebrate the mosquito and the diseases it communicated to French and English invaders. Yellow fever, like malaria, was an obstacle to European colonization of Africa. Europeans were responsible for transmitting it to the Americas and suffered themselves from the disease, as in the case of French soldiers who took part in Napoleon III's ill-fated expedition to Mexico. Yellow fever still exists in West Africa to the point that practically all adults are immune to it. The immunity of those who have survived the disease is solid and lifelong, producing an excellent immunity of populations. After the first American epidemic in Yucatán in 1648, others occurred. In 1793, 15 percent of the population of Philadelphia died. In 1905, in New Orleans and other American ports, five thousand cases were reported, resulting in one thousand deaths. In 1948 and 1957, there were numerous epidemics in Central America. Naturally, Africa was not spared; from 1962 to 1964, in southern Ethiopia there were one hundred thousand cases, with thirty thousand deaths. From 1978 to 1980, there were attacks in Brazil, Bolivia, Ecuador, Peru, Venezuela, Nigeria, Ghana, Senegal, and Gambia. In 1983, Burkina Faso and Ghana were also struck.

Effective vaccination began in 1927, but efforts to eradicate the virus failed and will remain unsuccessful because the pathogen can be sheltered by various hosts, including ticks.

## AN EPIDEMIC IS NOT CAUSED BY A MICROBE ALONE

In 1996, Laurie Garrett received the Pulitzer Prize for her study on new infectious diseases. In chapter 13 of her book, *The Coming Plague,* she compares the weight of a bacterium to that of a whale, and although the marine mammal is billions of times heavier than the bacterium, the

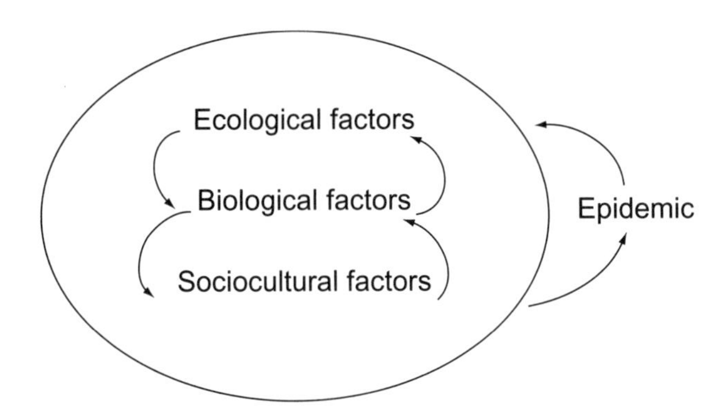


**FIGURE 1-1** The first loop of epidemics: an epidemic involves the emergence of biological, ecological (biotic and/or abiotic), and sociocultural phenomena.

author notes that the latter can kill the former. In an analogous way, microorganisms have been called the "wild beasts" of the future. In the case of epidemics, this way of seeing things is exaggerated insofar as a microbe does not suffice to create an epidemic. An epidemic is not caused by a pathogenic agent alone; it has non-biological causes as well. Sociological, cultural, and even abiotic factors such as floods and earthquakes play a role in the emergence of an epidemic, which itself in turn promotes the emergence of social phenomena and often new biological characteristics, as well as unprecedented psychological, cultural, and religious phenomena.

We can imagine the processes that give rise to epidemics as a series of loops. The first loop includes the biosociocultural and ecological aspects of epidemics (see Figure 1-1).

In large-scale epidemiology, it is impossible to separate the biological from the social, and it is therefore out of the question, when dealing with infectious diseases, to take a single approach, one that is either exclusively sociopsychological or limited strictly to the biological.

### *The biology of epidemics*

The biological aspects of the rise of epidemics are, of course, a reality that cannot be ignored. They take into consideration two phenomena: the virulence of the pathogens and the receptivity of the hosts. The aggressivity of the former and the resistance of the latter explain the almost regular oscillations in their accurrence that mark the complex events we call epidemics. The plague outbreak during the reign of Justinian began in A.D. 542 and continued to rage in a more or less acute form until the end of the seventh century, reappearing in 1348 and again leaving Europe in the eighteenth century. Influenza, which has a shorter periodicity, gives rise to pandemics or great epidemics every ten to fifteen years. This amounts to a situation of unstable equilibrium that is essentially biological in nature and reflects a coevolution or reciprocal adaptation between the virulence of the aggressors and the state of populations' immunity. The immunity of populations arises from its dynamic relationship, which is in a way a power relationship, with microorganisms. Sometimes the biological resistance of a human group is weak, poorly adapted to the biological virulence of the aggressor. It is generally agreed that Amerindians' receptivity to European germs was high, especially since these populations had probably been spared contact with them up to that point. Thus the Aztecs are thought not to have experienced any major epidemics before the Spanish conquest. More recent studies, carried out in connection with vaccination programs, have shown that among Amerindian children in North America who have been vaccinated against a bacterium that causes repeated infections, the protection obtained may be weak, as in the case of Apache children or Eskimo children. Human biological characteristics have been associated with weaknesses in protection against viral diseases such as hepatitis B and measles.

However, receptivity alone does not suffice to explain an epidemic. Even in groups that are receptive or susceptible, a "plague" is usually preceded by a stage in which the microbe amplifies—that is, by a phenomenon that increases the microbial population. The factors in this amplification are environmental and social; among them are health-care structures (hospitals, dispensaries) and confined concentrations of people (armies, slums, camps). Then the microorganisms in question often express a greater virulence in their infectious capacity. Such changes are commonly mentioned to explain the tidal wave of the Black Death epidemic. It has been shown how a simple mutation of a gene in *Yersinia pestis* was probably an important element that transformed the bacterium into a germ capable of producing an epidemic. Similar observations have been reported for other bacteria. On the biological level, therefore, an amplification of the microorganism's population and an increase in its virulence are often observed before an epidemic, the two being connected since the microorganisms are more numerous and more harmful.

As we have seen in the preceding pages, there are good reasons for thinking that, like the plague bacillus, the cholera vibrio and the flu virus undergo permanent modifications that are accompanied by an increase in their virulence. This is not so certain for the smallpox virus, whose original genetic constitution seems to have been able to produce an unchanging but powerful virulence. Microorganisms' propensity for modifying their genes is sometimes aided by human activities. In China, for instance, agricultural practices favor close contacts between animals of different species that carry the flu virus, among them ducks and pigs. To these spontaneous exchanges of viral genes among animals is added the use of animal and human excrement as fertilizer, thus constituting the most favorable mixture for producing transformed human viruses. The use of human excrement as fertilizer is a very ancient habit that is still practiced in many parts of the world and that favors the contamination of populations.

### *The "social" aspect of epidemics*

It is certain that sociocultural factors can facilitate the emergence of epidemics. Political crises and wars lead to the disintegration of administrative structures as well as to movements of human groups (armies, refugees) that are instrumental in the propagation and dissemination of diseases. Such conditions were met during the Hundred Years' War; they also explain many of the epidemics of cholera during the recent conflicts in Uganda, Rwanda, the Democratic Republic of Congo (formerly Zaire) and the like.

Like the movements of human groups, the transportation of merchandise is an element in the diffusion of pathogens. Histories of plague often begin with the arrival of contaminated ships in ports. In Europe, plague was a disease of ports and commercial routes, and that is why during the fourteenth-century epidemic, it moved along the trading route from Marseilles (a port on the Mediterranean) to Bordeaux, La Rochelle, and Bayonne (ports on the Atlantic). The slave trade was responsible for the spread of yellow fever to the Americas, and now international and intercontinental exchange in goods and travelers is so easy and important that it would take only a week or two for a flu epidemic to spread worldwide. It took the Black Death twenty years to spread from Mongolia to Bordeaux. Today, the average speed of a pathogen's propagation is infinitely greater. Cholera is supposed to have been transported in a few weeks from Bangladesh to Peru in the ballast of a ship. Depending on the sanitary conditions in the countries concerned, which are usually related to those countries' wealth and social organization, the disease will be propagated more or less rapidly. A good example of an epidemic related to social disorder is the waves of diphtheria in the former USSR. Until 1970, the disease was infrequent, but it began increasing in 1980. The mortality rate ranged from 2–3 percent in Russia and Ukraine to as much as 20 percent in Georgia, Azerbaijan, and Turkmenistan. Let us recall that here we are talking about the end of the twentieth century and of a country considered rich and developed.

Poverty and major social inequalities can incontestably favor the emergence of infectious diseases. People can be poor and deprived not only in poor countries, but also in New York, London, or Paris. We shall have occasion to emphasize once again the serious problem of the connection between poverty and diseases caused by microorganisms.

### *The ecology of epidemics*

Ecological phenomena can also play a role in the emergence of epidemics. Plague outbreaks increase during warm, humid periods, then decrease with the return of cold weather. In his *Times of Feast, Times of Famine: A History of Climate since the Year 1000,* Emile Le Roy Ladurie notes that climatic conditions (humidity, rain) fostered the emergence of the Black Death epidemic. The climate may have aggravated already unbalanced economic conditions and/or influenced the aggressivity of the pathogen. Cholera is a disease related to flooding, tidal waves, and other exceptional climatic conditions such as those caused by *El Niño*. Then all the conditions for the emergence of the disease are fulfilled, especially if the countries concerned are experiencing social chaos, migrations of human groups, poverty, and so on.

### *The consequences of epidemics*

The consequences of the dramatic events we have just mentioned are certainly immunological in nature, since the survivors of epidemics of plague or smallpox were protected, genuinely vaccinated.[27] This idea is not new; already in the twelfth century the Islamic philosopher Averroes noted that the smallpox virus never struck the same individual twice. We often see a simultaneous evolution of a germ's virulence and certain genetic traits of the human groups affected. In eighteenth-century Surinam, for instance, Dutch colonists experienced an epidemic of typhoid and yellow fever. Mortality rates reached 60 percent. A study of hereditary characteristics shows that descendents of these colonists have genetic characteristics that limit the extent of epidemics.[28] Taking into account the possible coevolution of host and parasite, natural selection should regularly increase the virulence of the pathogen, since the immunity of populations tends to reinforce itself. This is not always the case, as the example of myxomatosis (a viral disease fatal to rabbits) shows.[29] Syphilis may also have undergone a process of coevolution leading to the persistence in the patient of the pathogen, a spirochete, in a less virulent form. In fact, in the sixteenth century syphilis was a very serious disease that produced fatal internal and skin symptoms. Did the virulent form of the microorganism disappear along with its victims? Did the spirochete mutate into a less virulent form? History does not tell us.

Alongside these biological manifestations, which are certainly very important, there are others that are of great significance for the societies affected and whose effects can still be seen today. These effects are demographic, economic, social, political, psychological, religious, and cultural in nature.

### *Demographic repercussions*

The demographic repercussions of epidemics are obvious, but we need to be aware of the real impact that these ravages, even the ancient ones, had on cities, provinces, nations, and continents.

Let us take Easter Island as an example. In 1862, Peruvians landed on the island and deported a thousand natives to the mainland, where they were forced to work under inhuman conditions gathering guano. In 1863, under pressure from European governments, these poor wretches, only eighty-six of whom survived, were sent back to their country. Seventy-five of them died on the return voyage; the fifteen survivors were carrying various diseases, including smallpox. On arrival, they were welcomed with open arms, but by the end of the following year half the island's native people had died.

In the case of the Black Death, the depopulation of the European continent was exceptionally great. The disease carried off sixty thousand persons in Naples, two-thirds of the residents of Genoa, 80 percent of Majorcans, and at least half the inhabitants of Barcelona and Valencia. In Marseilles, Montpellier, and Avignon, it is estimated that 50 to 80 percent of the population died. In London, there were thirty-five thousand deaths among sixty thousand inhabitants. The frequency of deaths impressed contemporaries. In Paris, it is estimated that there were as many as eight hundred deaths a day, and in Vienne, six hundred. Epidemics certainly played a major role in the depopulation of France at the beginning of the seventeenth century.[30] For Amerindians of Central and South America, epidemics were synonymous with foreign invasions, terrible mortality, and the collapse of their societies. Tenochtitlán, which later became Mexico City, fell to Cortés on August 13, 1521. A year later, 50 percent of the city's inhabitants had died of disease. The Mexicans had to learn to live—or we should say, survive—with the epidemics that struck the country in 1531, 1541, 1564, 1576, and 1595. Thus it is estimated that the *Conquista* caused the death of fifty-six million Amerindians, for the most part as a result of epidemics. At the time, some Europeans considered the mortality rates among Amerindians as "normal" and as reflecting the will of God. For example, in 1548 the population of Santo Domingo was only five hundred, whereas earlier, at the time of the conquest, the island's population was one million. The governor of the island concluded that this showed that "God repented of having created a people so ugly, vile, and sinful." In 1630 the war between New England colonists and the Saugat Indians was accompanied by a smallpox epidemic that spared all the colonists but killed off large numbers of the natives. We can see why this selective mortality was attributed to divine will, whereas it was only the consequence of the immunological state of the two groups: one was immunized, one was not. The plague in Athens recounted by Thucydides was in large part responsible for the decline of that city. When the Hittites, an Indo-European people of central Anatolia (now part of Turkey), were invaded by the Egyptians in the fourteenth century B.C. the pharaoh's army brought with it the plague, which was fatal to local population.

All the diseases mentioned in this chapter have not had, at least not constantly, so deleterious an influence on demography, but it is nonetheless appropriate to discuss, along with the quantitative aspect of the demographic decline, its qualitative aspect, which in many respects had repercussions on the organization of society. For example, mortality rates within Catholic religious communities during the Black Death contributed to a spiritual emptiness that partially explains the metaphysical doubt among the surviving populations and, in a certain way, the future emergence of the Reformation. Sometimes terrorized populations preferred to rely entirely on the ecclesiastical authorities, as in Orvieto, which in 1354 named the pope and his legate as its lords for life, thereby abdicating its freedom. The death of priests increased the feeling of despair accompanying the great fear caused by the disease.

In the course of history, physicians, notaries, and gravediggers have also paid a heavy tribute to epidemics.

### *The great fears*

The way people behaved during the "historical" epidemics can enlighten us with regard to the events accompanying infectious diseases today. Ignorance about the origins of diseases explains a great deal about the earlier period. The fear generated by a disaster like a plague epidemic naturally led witnesses to believe in supernatural causes such as the conjunction of stars, a divine scourge punishing men's sins.[31] The most rational people of the time suggested that earthquakes were to blame, or perhaps air polluted by miasmas, whose corruption should be avoided.[32] Fear commonly leads to aggressivity, and popular belief, in past periods of confusion and distress, had an irrational need for scapegoats. It often happened that vengeance was taken on groups of men and women characterized by their difference. Even those who were tolerated in normal times were subject to the worst kinds of social rejection in times of crisis. Lepers, Jews, and gypsies paid a high price (sometimes in every sense of the term) during these troubled periods.[33] During plague epidemics, pogroms were common, and it was not until the the time of the Nazis that such persecution of Jews was seen again. At the time, neither Pope Clement VI nor the king of France was able to stop the massacres.[34] When groups could not be attacked, imaginary offenders were designated, people who were alleged to have plotted epidemics. It is probable that these accusations were often made for trivial, personal reasons, people seeking revenge in domestic or neighborhood disputes. After all, some people thought, getting rid of Jews was a good way to eliminate creditors, and often to confiscate their property. Those accused of spreading plague, who were said to mix pus from a bubo with animal grease and then spread it on door handles, were ferociously tortured before being killed in the cruelest ways (beating to death, breaking on the wheel, burning at the stake). The names of some of these alleged terrorists of the plague have come down to us through history.[35] Some of those tortured may have been looters, but were they willing vectors for the disease? After all, even the great sixteenth-century physician Ambroise Paré believed in the reality of these criminal activities.

Frightened people prefer to flee when they can, to seek refuge in safer regions. Montaigne was not the only representative of authority who left his city during an epidemic. Several kings of France—Charles V, François I, and Henry III—had to flee the plague. Nobles, municipal magistrates, judges, consuls, and even physicians often did the same. The common people had no resort other than to turn to the church and its miracle-working saints like St. Sebastian, St. Gregory, and St. Roch, who were supposed to be able to protect them against the disease.

### *Spirituality, culture . . .*

The town of Oberammergau in Bavaria shows one way that epidemics have influenced human behavior and that is still perceptible today.

This mountain town hoped to escape the 1633 plague by barricading its gates. But one resident violated the prohibition by returning home, without the knowledge of his fellow townsmen, after a trip outside the walls. The result of this disobedience was not long in coming and took the form of an epidemic that killed almost a hundred persons. The survivors, only too happy to have escaped death, decided to stage a representation of the Passion of Christ every ten years. The players were amateur actors, and the drama was such a success that it has since been staged hundreds of times. Today, seats are sold out long in advance. This venture, mixing religion, piety, and epidemic, is not the only evidence of the religious fervor that followed attacks by microorganisms. Processions of flagellants crossed Europe in every direction, seeking to expiate the world's sins by scourging their bare backs for thirty-three days as they passed through cities and villages. The authorities in the Roman Curia, worried by the zeal of these fanatical self-mutilators, excommunicated them. The worship of the Virgin Mary also underwent a revival as a result of Pope Gregory's apparent success in combating the disease by leading a procession carrying an image of the Virgin. The clergy also paid a high price during epidemics. In France, Henri de Belsunce, the archbishop of Marseilles in 1720, was so devoted to caring for the ill that he refused to leave the city; he became a local hero, and a major street in Marseilles was named after him. Many priests whose ministry obliged them to be close to the ill were less fortunate, losing their lives during epidemics. The resulting spiritual vacancy was not without consequences. The "plagues" in the history of humanity did not necessarily increase religious fervor and the moral connotation that usually accompanies it. Men and women sometimes abandoned their faith and occasionally sank into unprecedented licentiousness.

The impact of epidemics on culture also took numerous forms, from the replacement of Latin by vernacular tongues to the previously mentioned literary creations and pictorial, sculptural, and cinematic works.[36] In 1995, *Balto,* an American animated film by Simon Wells, told the story of the heroic exploits of a sled dog during a diphtheria epidemic in Alaska in 1925. This is a film version of the history of the real Balto, whose statue can be found in New York's Central Park near Sixty-seventh Street. The bronze statue honors the memory of the lead sled dog on the team that carried in record time, at temperatures of sixty degrees below zero centigrade, anti-diphtheria serum to twenty-five children in Nome, which was cut off from the rest of the world. The animated film and the statue are two very different artistic expressions associated with an epidemic. It is noteworthy that the great epidemic of Spanish flu has left practically no trace on culture. This is a special case, a kind of occulted epidemic, buried deep in the collective memory. The epidemics that accompanied the Spanish conquest of the Americas were responsible for another form of expression insofar as they led both to an extremely fertile creativity and to a truly extraordinary cultural collapse on the South American continent that resulted in the disappearance of peoples and their customs.

Among the aftereffects of the great historical epidemics the most important were surely those that affected demography with its social, political, and economic consequences, whether they were marked by social chaos, political erraticism, and general impoverishment or not. Periods of epidemics sometimes favored the emergence of new "national" structures. Thus the Bearn region of southwest France, a mountainous area, was spared the Black Death and took advantage of the disarray among its less fortunate neighbors in Aquitaine and the Pyrenees to obtain its independence in 1365.

In summary, plagues—understood in the broad sense of epidemic diseases in human history—sometimes followed abiotic disorders such as abnormal climatic conditions that may or may not have been associated with ecological repercussions. In addition to these environmental imbalances, which were often aggravated by human activities, demographic, sociological, and economic phenomena, as well as physiological and psychological phenomena, were also involved. Description of these events shows that not all of them are necessary conditions for the outbreak of an epidemic, but they participate in such outbreaks to varying degrees. The feedback loop represented in Figure 1-1 normally moves slowly, like a background noise combining human phenomena and the adventures of microorganisms. The mass of human groups and/or other species (rats, mosquitoes, fleas, squirrels, apes, etc.) plays a role in the buffering system, softening deleterious effects and avoiding an irreversible ratcheting up of the epidemic loop. It sometimes doesn't take much for this loop to be transformed into a ticking time bomb. A beat of a butterfly's wing, an element in the complex system that becomes corrupt or out of balance, and the ecosystem's fragile balance is destroyed in an irreversible and chaotic maelstrom. The models of historical epidemics described in the preceding pages suggest the necessity of a biosociocultural approach to this kind of event. Not only biological, but also abiotic, social, and cultural factors play a role in the outbreak of an epidemic, which in turn favors the emergence of other social, psychological, behavioral, cultural, and biological phenomena. In conclusion, let us note that at least some of the epidemics described above may appear to be "old stories about epidemics" that are a little dusty and perhaps even obsolete. To be sure, the models I have chosen to describe are exemplary, even caricatural, but to speak of these epidemics is not, in my view, to evince an outmoded attachment to the past—quite the contrary. To see this, one has only to consult on the Internet the most recent information provided by the WHO. For example, at the moment that I am writing these lines, the officials of this organization are expressing their concern about monkeypox, which usually affects people less than fifteen years old because they have never been vaccinated against smallpox. Should we have stopped vaccinating people against smallpox? Who can give a categorical answer to this question? Moreover, in 1997 the WHO noted epidemics of cholera in Somalia, Mozambique, Tanzania, Kenya, and Djibouti; epidemics of plague in Mozambique; and epidemics of yellow fever in Liberia. The WHO also recommends producing a flu vaccine combining all three types of influenza viruses. The calamitous triad of smallpox, plague, and cholera is still with us, but in a somewhat different form, insofar as the first of these "plagues" may have been replaced by monkeypox.

CHAPTER 2

# THE EMERGENCE OF NEW DISEASES

In 1994, Richard Preston, a teacher and researcher at Princeton University who also wrote for the *New Yorker*, published a book entitled *The Hot Zone, The Most Terrifying True Story You Will Ever Read* (Random House, 1994). The book was a great success, providing the basis for the film *Outbreak* starring Dustin Hoffman. However, the real protagonist of this sensational story is a frightening new virus: Ebola. Preston's scientific thriller seems to have opened new avenues for future best-sellers; Douglas Preston, Richard Preston's brother, later published in collaboration with Lincoln Child a work of the same kind, *Mount Dragon* (Forge Book/Tom Doherty, 1996). This science fiction novel deals with a lethal strain of the flu virus created through genetic engineering. It is no accident that fictional works about modern epidemics with a whiff of terror are being published. They testify, on the level of journalism, fiction, and popular literature, to a preoccupation that is a current reality in public health services: the emergence of so-called new germs.

## AIDS

Among the infectious diseases due to a germ perceived as new, acquired immune deficiency syndrome (AIDS) has a special place: it is the most social and political illness of our time. The fact that the AIDS virus is able to establish itself quietly among those it infects has helped it spread among human populations. It has taken on a worrisome global range. This new plague emphasizes more than ever that henceforth there will be one kind of medicine for wealthy people and another for the great majority of the world's population, which will never gain access to expensive therapies.

The first official mention of AIDS dates from June 5, 1981, when Atlanta's famous Centers for Disease Control and Prevention (CDC) reported five cases of pneumonia caused by *Pneumocystis carinii* in young male homosexuals. *Pneumocystis carinii* is a so-called opportunistic infectious agent that does not affect individuals with normal immunity, striking those whose defense mechanisms are deficient. On July 4, 1981, the *Morbidity and Mortality Weekly Report* mentioned the abnormally frequent presence of a tumor of the connective tissue below the skin, known as Kaposi's sarcoma, in twenty-six young men who were also homosexuals. At this time, Kaposi's sarcoma was known as the "gays' cancer." The "gay phase" in the emergence of AIDS is well known to the general public; it is magisterially described in Mirko Grmek's *Histoire du Sida* (Paris, 1989; *History of AIDS,* trans. R. Maulitz and J. Duffin, Princeton University Press, 1993).[1] Today, the disease is no longer limited to gay communities in the United States, Europe, and elsewhere, but has spread in a slow, inconspicuous, but terribly effective way to every continent. The spread of this scourge is a matter of particular concern in Africa and Asia, where AIDS is also a disease affecting children.

### *The origin and progress of the disease*

Since the discovery of the agent responsible for the disease, sensational scientific advances have been made in a relatively short time. Alongside work carried out in laboratories, methodical analyses by epidemiologists have allowed us to explain many things. This research, whose goal is to understand how pathological phenomena emerge, is so useful that a particularly successful epidemiologist, David Ho, was named *Time* magazine's Man of the Year for 1996 (December 30, 1996–January 7, 1997). The human immunodeficiency virus (HIV) was discovered in 1983 by Mme Francoise Barré-Sinoussi, a member of Professor Luc Montagnier's team at the Pasteur Institute in Paris, in a ganglion of an individual said at the time to be "at risk."

The historical antecedents of AIDS as we currently understand it are hardly older than 1979–1981. Is this modern form of the disease symptomatic of the emergence of a truly new virus, or is it just a different expression of a condition that existed earlier but manifested itself only in vague symptoms without a clinically discernible cause? AIDS, in addition to the symptoms that accompany it, is a pathology in the epidemiological, sociocultural, economic, and political context, and it is so peculiar that in any event it represents a new entity marked by its modernity. Infections by the virus involved may have existed earlier without physicians having understood what was the actual cause of the disease. Some twenty years before the emergence of the virus in 1980, troubling cases were uncovered, but they were not labeled in either the United States or Europe. It has even been claimed that the symptoms of AIDS in its current clinical form were described as early as 1952.[2] In short, there is every reason to think that it is the expression of the disease that is new, associated with sex, blood, and drugs.

The AIDS virus must have "vegetated" in Africa for a very long time, perhaps for centuries, before one or more events contributed to the pathogen's dissemination. It is reasonable to assume that the AIDS virus emerged from a close cousin that infected primates.[3] Did the virus pass from primates to men? Or from adapted, resistant individuals to others who were much more susceptible? Current conceptions tend to support the thesis that infectious agents come from the forests. While this has not been demonstrated for HIV, it is clearer in the case of the Oropouche virus.[4] The AIDS virus is not thought to have been transmitted by either a parasite vector or an insect. The emergence of the virus may have been favored by ecological disturbances such as deforestation, a practice that has long been customary in Africa, but which in recent decades has been greatly increased for commercial reasons. However that may be, everyone agrees that the AIDS virus is African in origin and very probably derived from primates; the African green monkey (*Cercopithecus aethiops sabaeus*) was the chief suspect. The most plausible theory is that of the "hunter." In Africa, chimpanzees are often considered prize game. They are regularly hunted and eaten. Now, many monkeys carry the simian $SIV_{cpz}$ virus, which can infect a person who has scratches or other lesions on his hands and comes in contact with the monkey's blood. $SIV_{cpz}$ is usually eliminated by human immunity, but it is possible that viruses adapted to the new host, producing a human immunodeficiency virus (HIV). This would explain the early variants of HIV-1 (as well as the transmission from primate to human). A recent study of 1,099 Cameroonians showed that the transmission of retroviruses, in this case simian foamy virus (SFV), from primates to humans was possible.[5] Research in molecular epidemiology indicates that HIV-1 evolved along with the chimpanzee *Pan troglodytes troglodytes* and that it had been present in this subspecies for centuries. This hypothesis was supported by a study of Marilyn, a female chimpanzee "resident" in an American animal house.[6] Its results are currently being debated by proponents[7] and critics.[8] There are in fact four notable forms of HIV: HIV-1M (the one involved in the pandemic), HIV-1O, HIV-1N, and HIV-2. More than thirty different species of monkeys have their own SIV. The chimpanzee is the origin of the three forms of HIV-1. HIV-2 is thought to derive from the sooty mangabey. The virus would have to have made at least five "jumps" to arrive at the current state of human contamination.[9] It is highly probable that HIV comes from a simian cousin that came in contact with our species in the course of hunting. We do not know, however, why this transmission occurred recently and not, for example, a century earlier, when monkeys were already being hunted in Africa.

Like that of syphilis, the history of AIDS will always remain associated with the idea of sex. The following examples show how important sex has been in the dissemination of the disease. In 1983, when AIDS appeared along the shores of Africa's Lake Victoria, it was called "Juliana's disease." The illness was spread through sexual intercourse between local women and merchants selling attractive clothing made under the "Juliana" brand. In 1984, a barmaid in the Tanganyikan town of Bukoba was directly or indirectly responsible for the transmission of Juliana's disease to twenty-four male and female inhabitants of the town. In Uganda, people spoke of the "slim disease" in a fishing village on Lake Victoria. This curious complaint was recognized as being AIDS brought there by pillaging members of the Tanzanian army or even by merchants or smugglers traveling back and forth between Tanzania and Uganda. For example, ten of the fifteen Tanzanian "merchants" tested at the time were HIV-positive. The sexual transmission of AIDS caused the road connecting Kinshasa and Pointe-Noire (Congo) with Mombasa (Kenya) to be nicknamed the "AIDS highway." The chief actors in the transmission of the virus were truckers, merchants, prostitutes, and, as is still often the case in these regions, armies on the march followed by their train of pillagers and preceded by refugees. The road from Malawi to Johannesburg and Durban is traveled essentially by trucks. It is known as the "Highway of Death," since 92 percent of the truckers have been infected by prostitutes in Durban or in towns along this highway.

At present, however, knowing these histories that mix sex, commerce, war, and pillaging does not allow us to say where AIDS began.

The history of the viral outbreak may (perhaps) no longer be very important, but the previously mentioned examples emphasize the role played by social disorder, wars, economic slumps, and poverty—all anarchical situations that can favor the emergence of an infectious disease.[10] Nevertheless, AIDS is not strictly speaking a contagious disease like measles, flu, or diphtheria. It is a transmissible disease that requires, in order to be contracted, contact or inoculation with contaminated fluids such as sperm, blood, vaginal secretions, and probably saliva, not to mention organ transplants. You don't get AIDS the way you get chicken pox.

AIDS is supposed to have migrated from Africa with Haitian laborers who went to work in Zaire in the context of cooperative agreements. Thus during the 1980s Haiti became a world center of high prevalence of the $V_1H_1$ virus. The island's political and economic difficulties had made Port-au-Prince one of the capitals of male prostitution and beloved of American "sexual tourists," who were to help spread the virus to their own country. We know what happened next; no one contests the role involuntarily played, in the ensuing development, by the lifestyles of American homosexual males and intravenous drug users. Drug use is another kind of epidemic that may precede AIDS.[11]

Currently, the great majority of contaminated individuals in the world were contaminated as a result of heterosexual relations. This is the case in Africa, where groups are often opposed to the use of condoms and where government authorities long prevented the dissemination of information relating to the extent of the virus's spread in their countries. This is all the more depressing because the economic and demographic consequences of a disease that attacks especially the young are considerable. Transmission that is essentially heterosexual more easily passes from males to females, and infections in the newborn are frequent. In Rwanda, the epidemic affects the population as a whole, and not only groups that are at risk. In many African countries, AIDS is the chief cause of death among young adults and an important factor in infantile mortality, the virus passing from the mother to the child during pregnancy. Will the death of the parents of young children transform Africa into a continent of orphans and thereby lead to a breakdown of society? A sad prospect for the cradle of humanity.[12] People's behavior in Africa helped establish an AIDS pandemic. One of the causes of the spread of this scourge is the social maelstrom with its cortege of disorder, poverty, famines, wars, fanaticism, and other reasons for the migration of populations, along with fear, despair, and resignation.

Africa does not have a monopoly on AIDS pandemics. Asian countries are also threatened by a third-world form of AIDS similar to that seen in Africa. Thailand is perceived as the country in which the sex industry is most flourishing, and the city of Chiangmai, while it is not the country's capital, is probably the world capital of trading in sexual pleasure. Five hundred thousand Thais make their living in the sex trade; prostitutes are often sold to shameless exploiters at a very young age by their parents. At least 70 percent of these young women are HIV-positive, whereas in 1989, only 4 percent were HIV-positive. Contamination has skyrocketed in this group of women, whose commerce is literally international. By 2010, mortality rates from AIDS in Thailand will have increased to

22 percent, while life expectancy will fall from seventy-five to forty-five years. Is it necessary to point out that in Thailand, as in Africa, there is strong opposition to the use of condoms?

India is also experiencing an explosion of HIV infection. Since 1993 there have been a million new cases per year, so that some ten million Indians were HIV-positive in the year 2000. Depending on the region, the spread of the disease in this immense country is connected with prostitution or with sharing the same needle in cocaine use. In the Manipur area, where there were many opium smokers, that old custom has been largely abandoned in favor of cocaine. The lack of syringes has resulted in HIV-positivity increasing from 1 percent to 80 percent of cocaine addicts.

At the end of the twentieth century, some thirty million people were infected with the AIDS virus (HIV) worldwide.[13] These included 530,000 in Europe, 860,000 in the United States and Canada, and 12,000 in Australia and New Zealand. The number of people who are HIV-positive is thus much smaller in rich countries, and these are, of course, the ones in which contaminated persons have access to the best medical care and the best currently available antiviral medicines, which are also the most expensive. In developing countries, HIV-positive persons have practically no access to appropriate medical care because of the cost of therapies.[14] For AIDS, a very special situation is developing on the global level, with one kind of AIDS in the Northern Hemisphere and another in the Southern Hemisphere, or an AIDS of rich countries and another AIDS of poor countries. Moreover, in Africa and Asia the virus is more likely to infect via the vagina, whereas in the Northern Hemisphere contamination basically has to pass through a breach in the vascular system. It is as if in the southern half of the planet the virus had adapted to heterosexuality, which is the most common mode of contamination in that area. Let us keep in mind that, in any case, social inequalities remain a constant co-factor in HIV infection.

AIDS appears as a supplementary factor that could be a determinant in the North-South imbalance. Western economic powers need the third world's natural resources, and any demographic crisis in Asia or Africa could have unfavorable consequences not only for the countries concerned but also for the rest of the world. It is not surprising that magazines like *Business Week* devote editorials to AIDS. Reports on the development of the illness in the third world that were presented at the Twelfth International Congress on AIDS held in Geneva in 1998, while encouraging in some respects, are disquieting in others. How can countries like Botswana and Zimbabwe offer young people, about 25 percent of whom are contaminated, treatments whose average cost is one thousand dollars per month? What will happen in central Africa, where whole areas are in danger of depopulation because of AIDS associated with other infectious diseases (see following)? How can we deal with a disease that contaminates sixteen thousand people a day, knowing that more than 90 percent of those infected live in developing countries?

The spread of this new disease, as we have seen, differs depending on the country, its culture, the nature of its society, its level of hygiene, and so forth. Even in a country like China, where rules of social and moral conduct are rigid and uniform, the number of those contaminated is reported to have doubled since the end of 1997 (an increase of some four hundred thousand persons). In mountainous southwest China the spread of the disease is connected with drug use, while in the maritime east it is connected with the growth of prostitution. In Thailand, on the contrary, although 2.3 percent of the population is infected, the number of new infections may be diminishing because of preventive efforts made in the sex trade.

In eastern Europe, it is drug addiction that is causing the spread of HIV infection, with 100,000 new cases estimated for 1997.

Recent therapeutic techniques are increasing the gap between rich countries, where death rates are falling, and poor countries, which cannot make use of new antiviral medicines. Treatment of opportunistic infections was already very costly, and in many third-world countries people who are HIV-positive are exposed to infectious agents that are particularly virulent (tuberculosis, for instance) or that are unknown in industrialized countries (malaria).

Once again, poverty, often combined with ignorance, amplifies microorganisms' deleterious impact. In western Europe, where prevention is encouraged, the number of new AIDS cases fell by 30 percent in 1997, and in the United States, where the number of new cases decreased in 1996, another decrease was expected in 1997. In the United States, there is a gap between better-off groups and the poor, especially African American and Hispanic people, among whom the rate of contamination has increased again.

Only expert physicians can decide whether to treat AIDS using current tri-therapy drugs. It is likely that in the future, tri-therapy will not only make use of chemotherapy but also take advantage of the auxiliary effects of an active immunotherapy known as "therapeutic vaccination" because it will not be merely a conventional preventive vaccination. At present, tri-therapy is able to improve the patient's condition to a considerable extent, but, unfortunately, forms of the virus resistant to these drugs are appearing.[15]

The AIDS pandemic continues to grow. In 2004 alone, three million people died and forty million HIV-positive individuals were alive and carrying the virus. Among the latter, five million were infected during 2004—or ten persons per minute. Africa, with the poorest countries in the world, is the continent most affected. In sixteen African countries, more than 10 percent of the population between the ages of fifteen and forty-nine is HIV-positive. In Botswana, 37.5 percent of the population is HIV-positive and in South Africa, 20.1 percent. The situation is serious throughout the sub-Saharan region, where three million people were contaminated in 2004, and women are especially at risk. However, progress has been made in some African countries. In Uganda, which has been hit particularly hard, efforts to prevent infection with HIV have been successful. In the 1980s, 15 percent of Uganda's population was infected, whereas in 2005, the rate had dropped to 6 percent. There is no question that the ABC approach based on abstinence, being faithful, and condom use has borne fruit, though it is not clear which of the three components of this approach is the most effective.[16]

In Asia and the Pacific, 1.2 million people were contaminated in 2004. To be sure, the percentage of individuals affected in heavily populated countries such as China and India remains relatively small but is nonetheless significant in absolute numbers. The scale of the epidemic in Asia depends more on the size of the population than on the virus's "explosive" (epidemic) spread. The disease spreads rapidly among groups at risk. In Asia, AIDS does not express itself as an illness affecting heterosexuals. The prevalence in Asia has never been more than 4 percent (this maximum was reached in Cambodia in 1999), and it has never been more than 1 percent during the prenatal period. In Asia, AIDS is mainly a disease affecting drug users, prostitutes, and homosexuals, but because of the size of the Chinese and Indian populations, this represents a large number of patients.[17] In the epilogue to this book, we will return to the role of denial in the course of epidemics. Here let us recall Bill Gates's "misadventure" in India. Gates wanted to help the country fight AIDS by making a large donation, which was received by the Indian prime minister and his aides. But Gates's description of the situation did not go down well; he was criticized by Indian authorities and represented as purveying misinformation.[18]

In eastern Europe and central Asia (Russia, Ukraine, the Baltic states, Moldavia, Kazakhstan) the epidemic is still growing. Its advance in Latin

America is equally obvious. In so-called rich countries, heterosexual contamination has become frequent, and although the rate of infection among homosexual males was declining, it appears that it is rising again. The progress made in treating AIDS seems to have led to a relaxation of private practices intended to prevent transmission of the disease. While AIDS treatments do not cure patients, it does allow them a life that is, all things considered, relatively comfortable. The treatment is daily and generally combines three drugs in order to prevent, so far as this is possible, the development of resistance. In disadvantaged countries, in order to make it easier to take the drugs and to ensure that they are taken regularly, they are combined in a single pill or tablet. Such combinations of generic drugs exist in poor countries, and include AZT, 3TC, D4T, and NVP.[19]

### *Biological resistance*

In early 1993, *Time* magazine published an article entitled "Are some people immune to AIDS?" The reporter referred to the case of a San Francisco artist who had been HIV-positive for fourteen years and had suffered only minor problems. Although not "cured," the patient did not present major symptoms even though he was infected with HIV. Examples of this kind are explained by the ability of these persons' immune systems to generate more efficacious immune processes to slow the virus's advance. Unfortunately, there are also other individuals who are more susceptible to the virus and in whom it is likely to advance more quickly. The search for biological markers associated with resistance or susceptibility to HIV's deleterious effects has been focused on the human leukocyte antigens (HLA) system, one of our biological entities that is often involved in the effects of microorganisms that threaten humans. Among the groups studied were the prostitutes of Nairobi, who long fascinated physicians and biologists because of their ability to have sexual intercourse with so many HIV-positive men without contracting AIDS.[20] It was observed that in those who had the HLA trait A2/6802, HIV-positivity was less frequent.[21] A later study of contaminated pregnant women showed that perinatal transmission of the virus from mother to child was less frequent in those who had the same HLA trait. Nonetheless, this protection was not effective against contamination while the mother was nursing.[22] Among the multitude of existing HLA traits, we see only a small number that are associated with resistance; for example, HLA A29 and HLA B57. Others, such as HLA B35, are known to confer greater fragility in the case of AIDS. On the other hand, it has been shown that there are individuals who are completely protected against AIDS, truly

TABLE 1 FREQUENCY OF MUTATION 32 OF CCR5 IN HUMAN GROUPS (AFTER C. J. DUNCAN).

| *Population* | *Mutated allele (%)* | *Normal genotype (+/+)* | *Heterozygotic genotype* | *Homozygotic genotype (−/−)* |
|---|---|---|---|---|
| Caucasians (Europe) | 10 | 81 | 1 | 18 |
| Caucasians (America) | 11.1 | 79 | 1.2 | 19.7 |
| Afro-Americans | 1.7 | 96.6 | 0 | 3.3 |
| Amerindians, Africans, Asians | 0 | 100 | 0 | 0 |

impermeable to the intrusion of the virus, which cannot find the key to open the door to their CD4 cells. Two cellular receptors are necessary for the infection: CD4 (complex of differentiation 4) and CCR5 (chemokine receptor 5).[23] There are mutations of CCR5 that prevent viral penetration.[24] Having one mutated allele provides partial protection, and having two provides a bulletproof anti-HIV vest.[25] Resistance to HIV is thus genetically determined via a mutation of the gene for the CCR5 receptor. The latter is absent in Africa; it appeared among humans who had already emigrated out of Africa some seven to ten centuries earlier. The mutated allele is relatively frequent.

Thus almost one European in ten is protected against AIDS. This advantage is ancient, and given the recent appearance of HIV, the current situation can be explained only by hypothesizing a process of coevolution in which the microbes and their hosts constantly adapt to each other in such a way that the relationship between them remains essentially unchanged. This is sometimes called a "Red Queen" coevolution, the reference being to Lewis Carroll's *Alice in Wonderland:* "'Well, in our country' said Alice, still panting a little, 'you'd generally get to somewhere else if you ran very fast for a long time as we've been doing.' 'A slow sort of country!' said the Queen. 'Now, here, you see, it takes all the running you can do, to keep in the same place. If you want to get somewhere else, you must run at least twice as fast as that.'" In some way, carriers of the favorable mutation (known as 32) benefit from a Red Queen coevolution resulting from the immunity of populations that earlier had to deal with another microbe. Two hypotheses have been proposed: carriers of the mutation selected by plague,[26] or else by smallpox.[27] (Their environment moves as fast as Alice and the Red Queen do, with the result that they seem not to have moved at all.

Thus with regard to the immunity of populations carriers of the CCR5-32 mutation, especially if the later is homozygotic, are essentially resilient with regard to AIDS. This privilege is the result of a selective process caused by a long coevolution of the plague pathogen, the smallpox pathogen, or perhaps both.

There is as yet no vaccine that can modify the immunity of populations to AIDS. Moreover, among persons whose immune systems are more or less seriously deficient, the use of live vaccines is out of the question, and in any event vaccination would have little effect if the immune system collapsed. The best treatment is, of course, prevention based on the use of condoms, blood products not contaminated by HIV, and needles and syringes that are sterile and used only once.

The use of sterile tools for injections seems to be a real problem. Mistakes in this area may explain the anomalies observed in some epidemiological studies. For instance, if it is claimed that sex alone is responsible for the explosion of HIV in Africa, how can we explain the fact that in Zimbabwe, during the 1980s the number of HIV-positive individuals increased by 12 percent per year, whereas the number of those contracting sexually transmitted diseases decreased by 25 percent? How can we explain the cases of HIV-positive children born of mothers who are HIV-negative?[28]

### *The future of AIDS*

We see, then, that our epidemiological future is a matter of concern, since a vaccine against AIDS has still not been made available or even really found. If we further reflect that HIV has a terrible tendency to mutate[29] and thus to develop new defenses against immunity or therapies, we can see that we have not yet managed to find a way to fight the pandemic and halt the worldwide advance of HIV.

Once again, HIV alone was not sufficient to cause the pandemic. It had to be associated with ecological, cultural, and sociological factors. Sex, drugs, wars, travel, blood transfusions and organ transplants, and the policies of government officials have helped spread the virus worldwide. On each continent, the virus and the disease have acquired specific characteristics that depend not only on biology (the type of virus, the ethnic origin of the patient) but also on cultural factors. Thus, contrary to Europe or North America, in Africa AIDS is basically communicated by heterosexual activity and affects men as well as women.[30] Since "groups at risk" play a minor role, the disease is more easily transmitted and advances more rapidly, and it is often accompanied by serious opportunistic infections and by cancer. In this particular case, the nature of the African virus matters, but the environment is just as important.[31]

Despite all efforts, the AIDS virus has continued to spread throughout the world. Between 1981 and 2004, AIDS killed twenty million people. At the end of 2004, women represented 47 percent of HIV carriers worldwide, but they were in the majority (57 percent) in sub-Saharan Africa. In 2003, young people between seventeen and twenty-four years of age constituted half the new cases of contamination. About five million poor victims of the disease were not able to buy the drugs necessary for their treatment.

Both sociologically and culturally, women are more vulnerable to AIDS because in most if not all poor countries, they are economically, socially, and culturally subordinate to their male sexual partners. They are especially at risk of contamination in countries where it is customary for men to have several sexual partners. AIDS has taken a very heavy toll on African women.[32] Women are in danger first of all because the probability of infection is for them very high. A woman who engages in unprotected sexual intercourse with an HIV-positive partner is twice as likely to be infected as a man who has unprotected intercourse with a woman who is HIV-positive. Thus worldwide women represent more than 60 percent of HIV-positive individuals between ages fifteen and twenty-four; sometimes they are orphans whose mothers died of AIDS. In addition to the biological inequality of the risk of contamination, it is very difficult for most third-world women to protect themselves against the virus—for several reasons. First, there exists the inequality of the sexes in patriarchal societies, where women are forced to accept a husband who is a notorious sexual nomad.[33] Second, in such cases a wife usually has great difficulty in getting her husband to use condoms. Third, women who are abused, economically dependent, and without resources often become prostitutes. In Africa, poverty is the chief cause of prostitution, and a woman forced to sell herself seldom refuses to have unprotected intercourse. Finally, in some African countries, such as South Africa, there is a high rate of rape because it is believed that an HIV-positive man who has intercourse with a virgin will be cured. All this indicates that under such conditions, the children of these women will be added to the long list of the HIV-positive. Many of them will not reach adulthood. Africa, the cradle of humanity, is becoming a continent of orphans. In 2003, worldwide fifteen million children younger than eighteen years old were orphans, and 80 percent of them lived in sub-Saharan Africa. It is estimated that in 2010, eighteen million African children will have lost at least one of their parents.

The Ebola virus is a horror among horrors, the champion of viruses responsible for hemorrhagic fevers such as Lassa and Marburg. Producing acute and quick-developing symptoms, this killer is so terrible that it could not produce a pandemic, because death rates are so high that the virus wipes out the hosts that provide its subsistence. Drawing on medical reports, Richard Preston describes in his best-selling novel the death of a patient who had been contaminated perhaps two weeks earlier.

> He becomes dizzy and utterly weak, and his spine goes limp and nerveless and he loses all sense of balance. The room is turning around and around. He is going into shock. He leans over, head on his knees, and brings up an incredible quantity of blood from his stomach and spills it onto the floor with a gasping groan. He loses consciousness and pitches forward onto the floor. The only sound is a choking in his throat as he continues to vomit blood and black matter while unconscious. Then comes a sound like a bedsheet being torn in half, which is the sound of his bowels opening and venting blood from the anus. The blood is mixed with intestinal lining. He has sloughed his gut. The linings of his intestines have come off and are being expelled along with huge amounts of blood. Monet has crashed and is bleeding out.
>
> The other patients in the waiting room stand up and move away from the man on the floor, calling for a doctor. Pools of blood spread out around him, enlarging rapidly. Having destroyed its host, the hot agent is now coming out of every orifice, and is "trying" to find a new host.[34]

We do not know where this new plague came from, but it certainly first emerged in Zaire in 1976. However, no one has been able to determine where the virus hid out between its attacks, that is, what served as its reservoir. It is likely that the Ebola virus leaves its natural habitat following ecological disturbances involving humans. This is a notion we have already mentioned, and which was notably emphasized by René Dubos at the end of the 1950s in his book *The Mirage of Health*.[35] The Ebola virus owes its name to the river in Zaire near which it was discovered. It is a filovirus,[36] and has three subtypes named for the countries in which they have produced epidemics among humans (Zaire Ebola, Sudan Ebola, and Côte d'Ivoire Ebola). The fourth subtype affects only primates, and is called Reston Ebola, after the city in Virginia that was the location of the pet store where a now well-known animal epidemic began.

### *Ebola's travels in Africa and beyond*

The first two epidemics of human Ebola occurred in 1976, the first in Zaire and the second in western Sudan. In these epidemics, 550 persons were infected, and 340 of them died. The third epidemic, in Sudan, affected thirty-four persons and resulted in twenty-two deaths. Between 1994 and 1996, there were three other outbreaks, including one in Kikwit (Zaire) with 244 deaths among 316 persons infected. The virus is communicated in situations of overcrowding, from human to human. Transported by tiny droplets of biological fluids (saliva, tears, blood), it can pass through the mucous membranes. However, this notion is very theoretical with respect to the realities on the ground, since in the previously mentioned epidemics the virus was frequently transmitted in hospitals. Let us note that the Ebola acquired in hospitals in the African bush developed in a very special kind of economic situation. The story has been clearly told by Laurie Garret in her book *The Coming Plague: Newly Emerging Diseases in a World Out of Balance.*

At the Catholic mission hospital in Yambuku, Zaire (now Congo), a patient turned up with symptoms that looked like an attack of malaria. The Belgian nuns treated him accordingly, and this resulted in the transmission of the Ebola virus to other patients. In this hospital, as in all the others in the country, the ridiculously low operating budget did not make it possible to buy enough syringes for the number of patients. Syringes and needles were reused several times, with the result that pathogens were transmitted within the hospital itself. Here we find the main cause of the spread of the illness: poverty, associated with a second, equally common cause, general ignorance and in this specific case, professional ignorance, since most of the nuns, like their lay assistants, did not have nursing degrees. Many of these devout women paid with their lives for their encounter with the terrible virus.[37] In Maridi, Sudan, the disease was communicated in the hospital under the same conditions, beginning with a sick person who came from N'zara. The emergence of Ebola in South Africa in November 1996 proved that the disease could leave central Africa and move to other areas. What happened was that a patient from Libreville, Gabon, came to Johannesburg's Morningside Medi-Clinic to be examined at the time that an epidemic of Ebola was raging in Gabon.

The peregrinations of Reston Ebola show how a virus can quickly spread throughout the world. The consequences that might have ensued had this virus been even slightly pathogenic for humans makes epidemiologists' blood run cold. The travels of Reston Ebola began on October 21, 1989, at the Fertile Company in Manila, where one hundred *Cynomolgus* monkeys were put on a KLM flight to Amsterdam. The animals were handled by the company's employees as well as by KLM employees at the airport. No one touched the animals during stopovers at Bangkok and Dubai, but in Amsterdam the cages were moved manually from the airplane to the airport's animal depot, which received some twenty-three thousand primates a year. During this stop, the Philippine monkeys more or less rubbed shoulders with other simians that had come from Africa and were being sent on to Mexico and Moscow. Afterward, the primates were put on a flight to New York, where they stayed in an animal depot at John F. Kennedy International Airport that received about fifty-thousand animals a month. On October 24, three days after having left the Philippines, the monkeys arrived at the Hazelton Primate Center in Reston, Virginia. By early December, fifty of the animals were dead. The remaining monkeys from the Philippine shipment and 250 others in the primate center were euthanized to stop the epidemic, which fortunately posed no danger to humans. A retrospective investigation showed that at least 173 persons had been in direct contact with the monkeys. It proved impossible to obtain any information about the area in which the monkeys had been captured, which was then controlled by armed rebels fighting the Philippine government.

In speaking of Ebola, it is usual to emphasize that because of the rapidity with which the virus strikes its victims, very often killing them, the spread of the disease is limited. That was true for the initial African epidemics, but not for the outbreak in South Africa or for Reston Ebola, whose trip from the Philippines to Virginia via Amsterdam took only three short days. Since this alarming episode, American health authorities have required that primates be quarantined for six weeks before being allowed into the country. Primates are accused of harboring the Ebola virus even more than HIV, but these animals, so valuable for North American and European scientific experimenters, are not the virus's usual reservoir—which no one has yet been able to discover. Probably the reservoir in question is made up of a species to which the virus is very well adapted, considering the stability of its genetic material (between the 1976 epidemic and the 1995 Zaire epidemic, a mutation rate of only 1.5 percent has been observed, very low under such circumstances), whereas HIV is constantly changing in order to defeat the host's defense systems. We assume that primates, long hunted for food or sale, temporarily left their natural habitat, somehow came in contact with the unknown reservoir, picked up the Ebola virus, and then returned to their natural habitat. The problem of Ebola's persistence has even been the subject of detailed articles in the American press, illustrating preoccupations that used to be limited to ecologists and defenders of animals but are now matters of general public concern.[38]

## MARBURG FEVER AND THE SALE OF PRIMATES TO LABORATORIES

The history of the first filovirus to show up in humans, the Marburg virus, illustrates the role played in the spread of the disease by primates that are subjected to scientific experimentation. Marburg fever is named after the German city where it first appeared in the world of *Homo sapiens*, in August 1967. Over a period of three days, the municipal hospital admitted three patients showing the same symptoms; all were employees of Hoechst, a multinational firm specializing in the production of vaccines. The company was at that time importing green monkeys for use in creating vaccines. It was soon seen that these patients, who were at first thought to be suffering from flu, had an unknown disease, a hemorrhagic fever. In the course of the following month, the hospital admitted twenty-three additional cases, all employed in the same industry. In Frankfurt, eighty miles away, six cases were diagnosed, including four among workers in a government laboratory that used primates from Uganda and the physician who had been treating these patients and was the head of the laboratory that carried out biological analyses on blood taken from the employees. Altogether, thirty-one persons fell ill, and seven died. The subsequent investigation showed that the primates had traveled from Entebbe, Uganda, to Belgrade and then on to Marburg and Frankfurt. Ninety-nine animals had been put on the plane in Uganda, and forty-nine were dead when they arrived in Belgrade. It turned out that they were infected before they left Africa. Since primates cannot be the reservoir for a virus so lethal to them, the reservoir—whether a mammal, an insect, or some other creature—remains to be discovered. This episode was, of course, very alarming for personnel of laboratories using primates. It resulted in the discovery of the Marburg virus, a previously unknown cousin of Ebola. Ebola and Marburg are not the only relatives of the family of viruses that cause hemorrhagic fevers, Lassa virus having also had notorious success in the category of extremely virulent microorganisms.

### *Lassa fever*

Lassa is a small town in Nigeria where the Church of the Brethren maintains a mission. In 1969, Laura Wine, a nurse at the mission, complained of pains in her joints, fever, and exhaustion. Her condition quickly deteriorated. Her body became covered with bright red pustules and the inflammation of her mouth and throat prevented her from eating. She was evacuated to the better-equipped hospital in Jos, a neighboring city. But nothing could be done for her, and after a few days of great suffering she died. Charlotte Shaw, a friend who also worked at the mission, had cared for Laura. She fell sick, and died just as rapidly. In the meantime, Dr. Jeannette Troup had been sent to investigate. She took the precaution of having Miss Shaw's serum analyzed in Columbia University's laboratory for tropical medicine. This made it possible to discover the virus, which was previously unknown. The chief nurse, Lily Pineo, who had cared for Charlotte Shaw, fell sick, showing the same symptoms as her unfortunate colleagues. She was transferred to New York, where she was put in intensive care, and recovered after losing fifteen pounds and all her hair. Dr. Casals, who had worked to isolate the virus, also got sick, but benefited from gammaglobulins produced by Lily Pineo, whereas a laboratory animal was less fortunate and did not survive the disease. Lassa fever reappeared in Jos in early 1970. On January 25, Dr. Troup carried out an autopsy and cut herself slightly through her surgical gloves. On February 18, she died. Lassa fever also appeared in Liberia, and Sierra Leone. In the latter country, it was noted that the rat *Mastomys natalensis* was particularly abundant in the villages that had been struck by the disease. The CDC's laboratories in Atlanta showed that the rat was the virus's reservoir. In fact, it was realized that, for a very long time, Lassa fever had been endemic in certain parts of Africa, and in some areas in Sierra Leone 40 percent of the villagers are immune to it, having antibodies against the Lassa virus that provide the basis for the immunity of their populations. Later on, the virus was found in the rat over much of Africa. It turned out that Africans have been in contact with the virus since ancient times and that it was the emergence of the disease within a hospital and among non-Africans that produced the appearance of an epidemic. Another particularity of this hemorrhagic fever is that the virus can be treated with a medicine, ribavirin, whose cost is, however, too high for use in most African countries. For some people, Lassa, Marburg, and Ebola represent the plagues of the future. Marburg and Ebola are

*Filoviridae*, whereas Lassa, like its American cousins Machupo and Junin, belongs to the family *Arenaviridae*.

## OTHER HEMORRHAGIC FEVERS AND ECOLOGY

In the introduction to this book, I briefly mentioned the emergence of Bolivian hemorrhagic fever after a program of agrarian reform was undertaken in Bolivia in the late 1950s. This program resulted in major deforestation, the forest ultimately being replaced by farms growing corn. The presence of this grain promoted the multiplication of a local mouse, *Calomys callosus,* especially since cats had sometimes been eliminated by the use of DDT. The proliferation of these mice, combined with the destruction of the forest, facilitated human contact with a possible virus reservoir. The Machupo virus, which was harmless to the mouse, spread by using the mouse as a vehicle close to places where the local people lived. The mouse disseminated the virus through its saliva, its urine, and its feces, and this was an effective way of contaminating everything the peasants might handle, drink, or eat. The virus of Bolivian hemorrhagic fever, or Machupo virus, belongs to the *Arenaviridae* and has an Argentinean cousin called Junin, which was able, under similar conditions (corn growing, use of herbicides), to find a host in the *Calomys* mouse and contaminate Argentinean farmers. In both cases, these fevers producing high morbidity and mortality rates struck rural areas and led to economic disturbances for populations with modest incomes and living conditions that were often precarious.

## THE ARRIVAL OF THE HANTAVIRUS, OR THE CONSEQUENCES OF RAINY WINTERS IN THE DESERT

During the summer of 1993, I had occasion to speak with my friend Fred Koster, a specialist in infectious diseases at the University of New Mexico and at the Bernalillo County Medical Center, the university hospital of the city of Albuquerque. Fred was not the only person on the banks of the Rio Grande who was worried about what was happening in the so-called "Four Corners" region where New Mexico, Arizona, Colorado, and Utah meet. In early May 1993, on a Navajo Indian reservation in that area, a young Navajo woman had rapidly died of an infection that at first seemed relatively banal but was then complicated by major respiratory difficulties. The young woman's boyfriend soon showed the same symptoms, and also died. Other people fell victim to the disease, and the small local hospital in Crown Point was soon completely overwhelmed, with the old Navajo chief Zah more worried than ever. Dr. Bruce Tempest of the Indian Health Service in Gallup immediately notified health authorities of five deaths due to this new disease in the Four Corners area. At first, the disease had a mortality rate of about 75 percent, but physicians working for the federal and state governments soon found improved ways of treating patients and diagnosing the disease in its early stages. After considering the possibility of plague, which still occurs on these reservations, investigators found that the respiratory distress was due to a then unknown hantavirus.[39] The pathogenic virus uses the deer mouse (*Peromyscus maniculatus*) as its host. The virus, which belongs to the Bunyaviridae family, has relatives on other continents that were already known at the time of the events in the Four Corners region. The disease in question is thus not really a new disease, since in retrospective studies of the tissue of patients that had earlier been treated for ill-defined complaints it was shown, with the help of laboratory work, that a patient had suffered from the previously mentioned syndrome fifteen years earlier, in 1978. Moreover, the local tribes' oral tradition, which constitutes the Indians' tribal memory, reports incidents of the same kind occurring in earlier times. The emergence in 1993 was ecological in origin. The winters of 1992 and 1993 were exceptionally rainy in New Mexico, and this favored the proliferation of insects and a heavy production of pinecones. The deer mouse that lives in this area proliferated, thanks to abundant food, and its population increased tenfold. Up to a dozen mice per acre were counted, a figure more than sufficient for these little mammals (which like to frequent human habitations) to infect the Indians. The period of the mouse overpopulation was brief: between May and July 1993 the number of *Calomys* mice fell from 12 to just one or two per acre; owls, foxes, and rattlesnakes saw to it that the earlier balance was reestablished.

While there is no doubt that the prime mover in the Four Corners epidemic[40] was an ecological imbalance, it is nonetheless true that the populations affected were poor and lived under precarious hygienic conditions that explain the contact with the rodents. Seventy percent of those who fell sick with the disease had been involved, directly or indirectly, in domestic tasks such as sweeping floors, which caused dust to fly about. They were contaminated by involuntary inhalation of microparticles infected by mouse excrement. This kind of thing was frequently seen during the Machupo epidemic in Bolivia. If Navajo women had had vacuum cleaners, would there have been an epidemic? In October and December 1996, epidemics of hantavirus pulmonary syndrome occurred in Chile and then in Argentina.

The peculiarity of these latter epidemics demonstrated a possible transmission of the virus from one human to another. Other cases were reported in Brazil, Canada, Uruguay, and Paraguay, making the hantavirus the agent of a pan-American disease shared by humans and animals.

### THE AMERICAN LEGION, OR HOW A VETERANS' MEETING GAVE ITS NAME TO A DISEASE

Legionnaires' disease is an example of an old disease brought to light by an exceptional situation connected with something that long symbolized the comfort of "the American way of life" and technology in the United States: air-conditioning. The disease was a matter of great concern at first but no longer frightens people. Although the causes of the disease are known, the clientele of hotels and other public places still sometimes find themselves inhaling the infamous bacteria.

In 1976 members of the American Legion of Pennsylvania gathered in one of the largest hotels in central Philadelphia, the Bellevue-Stratford. This gathering was planned for a long weekend, which members of this respectable association spent together in a single place, their hotel, where they met, ate, and slept. In all, there were 4,400 participants; by the third day of the meeting, 221 of them had symptoms of pneumonia. Thirty-three of them, who were no longer young, died. The bacterium responsible for their deaths, which was then unknown, has since been identified and named *Legionella pneumophila*. It had proliferated in the water used in the hotel's air-conditioning system. The latter produced a constant flow of air carrying *L. pneumophila* to the hotel's customers and also to Broad Street, where the hotel was located, producing "Broad Street pneumonia."[41] In 1974, in the same hotel, there had been an earlier, undiagnosed outbreak that affected twenty people attending a convention. Epidemiological investigation showed that the disease had occurred in a hospital in Washington, D.C., as early as 1965, and again in a hospital in Pontiac, Michigan, in 1968, where 144 employees and visitors were also affected. The affliction first called "Pontiac fever" was Legionnaires' disease. The bacterium responsible is associated with aqueous environments and multiplies easily in natural or artificial reservoirs. Natural reservoirs include mud, mountain streams, lakes, and warm springs. In the natural environment, the bacterium can feed on nutrients provided by algae, or better yet, take up residence inside amoebae, protozoa that become true Trojan horses in which the bacterium finds food and protection against antiseptics. In artificial environments created by humans, such as reservoirs in air-conditioning systems, *L. pneumophila* develops perfectly, with the temperature, the iron, the ad hoc nutritive environment, and the absence of competition with other microorganisms all contributing to the pathogenic bacterium's propagation. Aerosolization permits the bacterium's transmission to future victims. The latter are not all equally susceptible to the disease. During the epidemic in Philadelphia, elderly men, whose immune systems were probably weaker, were most affected. Smokers are more susceptible than nonsmokers; women are three times less often affected and children almost never. Even among those whose immune systems are not compromised, and even with treatment with antibiotics, this can be a serious disease (with a mortality rate up to 7 percent). Legionnaires' disease is a product of nature and industrial culture. The bacterium was transported by fresh mountain streams to the unnatural water reservoirs of pleasure palaces where it could corrupt the air breathed by a clientele confined in a comfortable but not always healthful atmosphere.

### LYME DISEASE

Lyme disease is connected with ecological changes brought about by human activity. It was first discovered in 1975. It is a strange disease that astonished two mothers in the little town of Lyme, Connecticut. They had noticed that some of their children had been feverish for several days and had pain in their joints. The children were initially thought to be suffering from juvenile rheumatoid arthritis, an immunological disease of the joints. This hypothesis was abandoned when the disease struck a dozen children in the town, an abnormally high frequency for rheumatoid arthritis in view of the epidemiological data and the fact that the form of arthritis in question is not transmissible. The disease appeared during the summer in a town that was a rural suburb, and this suggested that the origin was a parasite, and more precisely an arthropod. The difficulty was that the symptoms did not correspond to any known infection, whereas the disease affected a very limited area that in 1977 included neighboring towns: Lyme, Old Lyme, and East Haddam. At that time, the strange pathology had struck thirty-nine children and twelve adults. It turned out that in certain cases the affliction was accompanied by the presence of the deer tick, *Ixodes dammini*, now called *Ixodes scapularis*,[42] which was the carrier of the microorganism involved. The infectious agent was identified in 1981 by Willy Burgdorfer by dissecting the ticks to culture the microbe they were carrying. The microbe is a spirochete that has since been named *Borrelia burgdorferi*. The pathogen is thus a bacterium, not a virus. The number of cases of Lyme disease in the United States continued to increase, passing the five hundred mark in 1982 and the ten thousand mark in 1992. The disease is now found in Europe as well. It may look like flu at first, and then lead to meningitis, heart trouble, and rheumatism. It is found all over the world, wherever *Ixodes* ticks can carry *B. burgdorferi.* Lyme arthritis has emerged as a result of ecological changes brought about by human activities connected with the agricultural and industrial development that began shortly after the Pilgrims' arrival on the *Mayflower.* To be sure, the disease existed earlier, but remained unknown because it had never taken on the magnitude it assumed in the 1975 outbreak. The previously mentioned ecological disturbance is associated with humans' destruction of the forests in the northeastern United States. After the arrival of English colonists, Massachusetts became the continent's first metallurgical center. Iron was produced there, and this required a great deal of charcoal, which led to systematic logging of New England's forests.[43] Deforestation was also encouraged by the establishment of larger and larger farms. At the beginning of the nineteenth century, the region was practically denuded of forests, which led to the disappearance of the deer that liked to live at the forest's edge, and also to the disappearance of their natural predators—wolves, coyotes, and cougars. Toward the end of the nineteenth century, the metallurgical industry in the east declined; at the same time intensive agriculture moved toward the Midwest, while farms in the east declined; in number and size. Finally, local and federal governments gradually adopted policies intended to preserve the forests, which reappeared in the form of woodlots, smaller but scattered over the territory and surrounded by brush. The rehabilitation of the deer's habitat without the return of its predators resulted in the deer's proliferation and thus to a proliferation of the *Ixodes* tick infecting other mammals and, of course, humans.[44] It is appropriate to emphasize here that the deer's proliferation was such that some Americans call the animal a "rat with hooves." The deer population has become so large that it grazes calmly near houses and even in city parks. Contamination of humans is all the easier because people living in New England's cities, like everyone else in the country, want to live in the suburbs, combining the charms of the countryside with the pleasures of the city. These suburbs, which amount to cities in the countryside, facilitate contact between humans and the *Ixodes* tick through the intermediary, among others, of the "white-footed mouse," a host liked by the parasite. Lyme disease can be transmitted by other ticks as well; in Europe it has been transmitted by sheep ticks and in California by a parasite of the wood rat. Last but not least, the parasite can also be transported by birds.

Lyme disease illustrates a situation in which *Homo sapiens* has destroyed an ecological environment and caused the disappearance of forests, herbivorous mammals, and their carnivorous predators. The new situation lasted about two centuries, and then the convalescence—the spontaneous or instigated rehabilitation—of the affected territories began. The ecological restoration was not (and is not) a restitution of the earlier state; one imbalance was replaced by another in which three living species are dominant: the deer, the mouse, and the tick, whose digestive system still provides an incubator for *Borrelia burgdorferi*. Lyme disease is not fatal, but its repercussions on public health are significant, and its cost in the United States runs to billions of dollars.

### DISEASES CAUSED BY PRIONS. BOVINE SPONGIFORM ENCEPHALOPATHY (BSE): MAD COW OR HUMAN MADNESS?

The chronology presented here[45] summarizes the main events in the history of mad cow disease in Britain. It brings out the authorities' reluctance to recognize the harmful consequences for humans of a bovine disease caused by appalling livestock-raising practices that consisted in feeding herbivores products made from sheep carcasses. British officials are still dealing with the unfortunate story of mad cow disease. As early as 1990, the British minister of Agriculture decided that there was reason to be concerned; in the United Kingdom, thousands of cattle were suffering from BSE. Sales of beef continued to fall, and the European Community put an embargo on British beef. In the same year, Britain decided to create the Spongiform Encephalopathy Advisory Committee (SEAC), an assembly of experts that regularly informs and counsels the government regarding the BSE problem. The initial concern was economic: was BSE going to destroy the British beef industry? However, by 1994 it was understood that meat from sick cattle had been able to contaminate *Homo sapiens* and might again, causing a painful, distressing, irreversible malady, Creutzfeld-Jakob disease. Since 1996, there has been scarcely any doubt about the risk to humans, because in 1994 a "variant" of Creutzfeld-Jakob disease appeared, a form of the disease that struck young people and not only, as had usually been the case, people over sixty.[46] The new form of Creutzfeld-Jakob disease is caused by what French public health officials call an "Unconventional Transmissible Agent,"[47] more commonly known as a prion. The term "prion" is an abbreviation of "protineaceous infectious particle." So far as mad cow disease is concerned, the animals' problems arose from cattle raisers' feeding them sheep in the form of ovine proteins that had been industrially extracted from carcasses and then incorporated into cattle feed. In the late 1970s, a British factory changed the method of extracting the protein in order to increase the yield; the raw materials contained the remains of sheep that had suffered from scrapie.[48] The modification of the industrial process made it easier for the agent of scrapie, a prion, to pass into cattle feed. Later on, the contaminated cattle showed behavioral problems that caused them to be called "mad cows."

### *Prions*

It was shown that the brains of the sick animals contained prions. Prions are peculiar transmissible agents. Being neither viruses, nor bacteria, nor parasites, they do not have even have DNA or RNA that would show the presence of genetic material. A priori they lack all that, prions being nothing but proteins. BSE, Creutzfeld-Jakob disease, and related problems are now considered diseases caused by prions.[49] When Stanley Prusiner suggested that prions might be the transmissible agents, his hypothesis was met for the most part with a lukewarm reception among scientists. It was then a heretical notion that is still not universally accepted today and that shattered the classical canons of contagious or transmissible diseases. However, like BSE, Creutzfeld-Jakob disease is transmissible. Prion-related diseases can jump the species barrier. Creutzfeld-Jakob disease can be transmitted to chimpanzees, and BSE can produce Creutzfeld-Jakob disease in humans. In addition, we know of a hundred cases worldwide of iatrogenic (caused by medical treatment) transmission of Creutzfeld-Jakob disease—after grafts of corneas or dura mater, with the use of growth hormones taken from cadavers' brains, during intracerebral exploratory surgery, and perhaps even through blood transfusions. Another particularity worth mentioning with regard to transmissible prion-related diseases is that this "contamination" does not provoke any immune response. To be sure, there are individuals who are (or would be) genetically programmed to be susceptible to prions, but the phenomenon in question here is not in any way an immunological one. In this case, it is not the absence of a faculty of immune response that explains, as is usual, the organism's failure to react; the predisposition is biochemical and metabolic in nature. All the epidemics mentioned previously and those we will discuss later involve the natural means of defense, individual immunity, whereas in the case of prions the immune system completely ignores the buildup of the abnormal protein.[50] Although it is clear that prion-related illnesses represent a new form of epidemic that has nothing to do with immunity, in many other respects the resemblances are striking. Among the convergences that are worth emphasizing is the development of agriculture and livestock raising. We said that cows went mad from having eaten sheep, and it was indeed in order to increase meat production while at the same time lowering costs that someone had such a crazy idea. Who is the maddest here, humans or cows?

Technology applied to medicine helped promote the transmission of Creutzfeld-Jakob disease—in a limited number of cases, to be sure, but sometimes when the use of dangerous substances is not completely justifiable on medical grounds. Consider, for instance, the question of height. We live in a normative world in which being short is a disadvantage, and so contaminated growth hormones, deriving from the human thymus, may have sometimes been—who knows?—used more for aesthetic reasons than in order to correct a genuine infirmity. Thus, behavior influenced by fashions that are ultimately foolish may have promoted transmission of the disease. The diffusion of BSE was also encouraged by international trade. Contaminated animal feed was exported illegally from Britain to the rest of Europe and probably elsewhere. A month does not go by without our discovering that irresponsible merchants have gotten around the prohibitions and falsified the labels on forbidden meat in order to make it easier to sell fraudulently.

## SO-CALLED NEW DISEASES: WHAT ARE THEY? WHERE DO THEY COME FROM? WHERE ARE THEY TAKING US?

The so-called new infectious diseases that have been mentioned in this chapter are only a few examples among many other pathological entities connected with microorganisms that we now know about. Since 1973 more than thirty previously unknown pathogenic agents have been identified.[51] Among the ailments we have chosen to describe, most are viral in origin; two are bacterial diseases (Legionnaires' disease and Lyme disease). Lyme disease is caused by the proliferation of a parasite that is not limited to humans (*Ixodes scapularis*). The maladies discussed represent only a minority of the infectious diseases that have been identified since the 1970s. The examples chosen enable us to emphasize what has been observed during most of the events considered "new," namely, that the factors in these epidemics included both ecological (abiotic or anthropic) and biological elements and that the sociocultural circumstances often

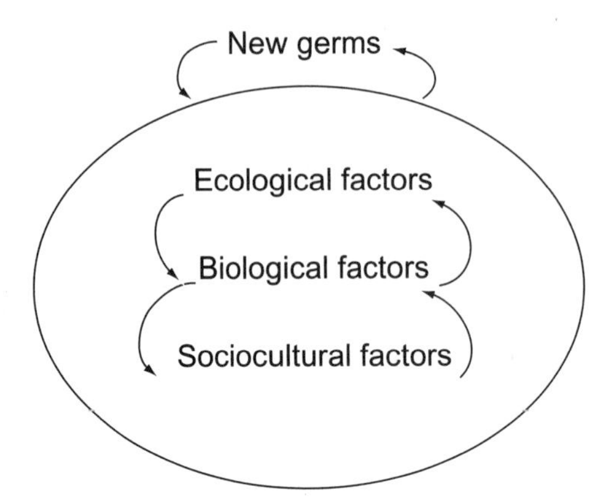


**FIGURE 2-1** The loop of new germs, or the second feedback loop of infectious diseases.

played a role. The previously mentioned factors are not necessarily associated with each other; they may interfere with each other, and often the outbreaks that they provoke have feedback effects on the generators of the disorder. These are not simple transfers from one process to another, but rather, as in the cases mentioned in the preceding chapter, interactions that lead to a second feedback loop (see Figure 2-1).

## ECOLOGY AND EMERGENT PLAGUES

### *Climate*

Schematically, it is possible to consider the ecological factors participating in the outbreak of epidemics or any other eruption of microorganisms as being of an abiotic and/or anthropic nature. Changes in climate can, via the occasionally major ecological modifications in which they result, have an effect on the equilibrium of an environment and encourage the proliferation of a pathogen. We have already mentioned the important role played by *El Niño* in certain outbreaks. This seems clearly demonstrated in the case of cholera (cf. Chapter 1), and it is in accord with the present discussion insofar as the cholera vibrio involved is a new form of that bacterium. *El Niño,* which has been implicated in many situations, is also thought to have been partly responsible for the hantavirus (or Sin Nombre virus) epidemic in the Four Corners region. In addition, it has been blamed for the abnormally humid winters in New Mexico in 1992 and 1993, but it has to be admitted that *El Niño*'s role is in this case more speculative than proven. Whether or not *El Niño* was involved, there is no question that the change in climate increased the deer mouse population, simply because the food supply, in the form of hazelnuts and pinecones, had been significantly increased by the wet weather. As the deer mouse population grew, the Sin Nombre virus grew along with it; and the same phenomenon has been observed in California.

Between November 1995 and March 1996, 540 cases of rheumatoid fever caused by the Ross River virus were reported in southwest Australia, particularly in the coastal region south of Perth and on the nearby islands. The disease is transmitted by *Aedes camptorhyncus,* and its abnormally high incidence coincided with a proliferation of this mosquito. The unusual numbers of the vector resulted from uncommonly high rainfall followed by equally high temperatures. Exceptionally high tides were also noted, along with the arrival of three hurricanes. Taken together, these uncommon ecological conditions are considered by epidemiologists to have been responsible for the outbreak.

### *Forest*

The ecological changes brought about by *Homo sapiens* are disparate, diverse, and numerous. Among these transformations of nature that can result in the proliferation of microorganisms are many that result from agricultural practices and the ways in which these practices evolve. The need to use the soil has led to the deforestation of certain areas. Deforestation played a role in the emergence of Lyme disease in New England. It has also been and continues to be a problem in Africa. René Dumont, for reasons other than ecological ones, has stressed the deleterious consequences of deforestation in Africa, a practice that is now exacerbated, but has long been customary among certain peoples on a continent where, as Pierre Gourou puts it, "the African farmer is an eater of forests."[52] Human societies, as soon as they reached a certain size, began to weigh on the environment of the soil that fed them. The invention of agriculture in the Neolithic Age led to a process of clearing the land and selecting certain species of plants that is still with us. For example, in Costa Rica 30 percent of the tropical forest has disappeared and been replaced by cultivated land. The extensive destruction of forests has modified the albedo, that is, the relation between the solar energy reflected by the earth's surface and the incident energy received. Areas that have been stripped bare by deforestation have an increased albedo. The worrisome current levels of deforestation could, in my view, promote the emergence of pathogens in two different ways: first, simply by facilitating contact between the pathogen in question and humans, and second, by modifying the relationships (the "biocenosis") among microorganisms, insects, mammals, and humans in the area concerned. This is in accord with the scientific historian Mirko D. Grmek's idea of "pathocenosis" as an encounter between demography and the biocenosis of microorganisms. As Grmek puts it, "the study of the distribution of diseases in terms of their frequency raises a problem that corresponds to that of the distribution of animal and vegetable species in relation to the individuals living in a biocenosis."

The preceding lines should not be taken to suggest that deforestation is synonymous with the evolution of agriculture. Not all deforestation is connected with farming. The tropical forest loses about twenty-seven million to thirty-nine million acres a year because of agricultural needs (about 50 percent), but it also loses ground to commercial logging of timber (26 percent), firewood cutting (14 percent), and clearing for livestock pastures (11 percent). Since the Neolithic period, agriculture and deforestation have been involved in the emergence of diseases. In the case of Africa, for example, it is likely that at some point in the distant past forests covered much of the continent. Farmers had no alternative but to cut down these forests, especially in West Africa, where anthropoid apes (pongids), among other species, were driven back. In this case, mosquitoes came in contact with humans and were able to pick up on the newly conquered areas pathogens that had been carried by apes. The proximity of the apes helped the simian pathogens' adaptation to humans. This was probably how malaria, which will be discussed later, contaminated humans. It is likely that the same process caused the emergence of new germs, the farmers and their families being exposed to insect bites. In addition, sedentariness is an element facilitating the establishment of diseases such as those caused by salmonella bacteria and roundworms. The recently seen outbreaks of diseases connected with deforestation (Oropouche, Lyme disease) are only modern versions of ancient processes. In India, the *Mahabharata* recounts the destruction of forests between the Ganges and the Jumna by Arian invaders around 2000 B.C. This episode might be involved in the appearance of a form of typhus due to a microorganism called *Rickettsia tsusugamuchi*. Alexander the Great's destruction of forests in 327 B.C. may have led to the disease's spread.

Forests are considered an extraordinary lode of living species. According to Wilson, there are 1.4 million known living species on earth, and 98.6 million that remain to be identified. The greatest number of living species is found in rain forests; for example, Wilson found forty-three different species of ants on a single tree in the Amazon forest. In November

1944, there was an epidemic of the Ebola virus that affected chimpanzees in the Tai forest in Côte d'Ivoire (Ivory Coast). This involved a new species of the simian Ebola virus. Its appearance after heavy rains suggests that an insect or some other arthropod was involved in the disease's transmission. It is also possible that the virus's host was some little mammal that the chimpanzee ate. Dr. Bernard Le Guenno of the Pasteur Institute was sent to inspect the site (some seventeen hundred square miles of forest), but despite his efforts he was unable to find the virus's natural host.

The destruction of forests can bring about an imbalance, a rupture of a harmony that is often fragile. In the 1980s, the ecologists Paul and Anna Ehrlich, who were patient observers of the forests, proposed the "rivet hypothesis" to explain the dangers of deforestation. They compared the ecosphere to an enormous airplane in which each rivet corresponds to a species. The loss of a few rivets, even though it does not greatly modify the airplane in general, can lead to its crash. The sylvan origin of many pathogens is widely acknowledged, and thus any attack on the forest can lead, in the long run, to the appearance of a new disease, in the sense defined above.

Deforestation is not synonymous with the evolution of agriculture, because trees are cut down independently of any agricultural goals. As such, agriculture modifies ecological niches or the biocenosis (including microorganisms, in this sense of the term).

### *Agriculture*

Intensive agriculture can influence ecology by diverting the natural flow of water in order to use it for irrigation. The Rift Valley fever described for the first time in 1930 has since unexpectedly reappeared in a dozen sites throughout the world.[53] The fever's vectors are the *Aedes* or *Culex* mosquitoes, which transmit the virus from sheep to humans. Among the modifications of the environment responsible for the proliferation of these mosquitoes, the construction of dams creating artificial reservoirs of irrigation water is particularly important. An analogous situation is seen in the case of many tropical diseases. We note that when agriculture encourages the emergence of a pathogenic germ, it does so by modifying the ecological conditions in such a way that *Homo sapiens* comes in contact with a new deposit of the pathogen. Contamination of humans is promoted on one hand by contact with the pathogen, and on the other by the relatively great number of microbes present. For example, the Hantaan virus causes the serious Korean hemorrhagic fever. This disease, which has long been known in Asia, strikes on average one hundred thousand Chinese peasants every year. The virus's favorite host is the field mouse, *Apodemus agrarius,* which contaminates humans during the rice harvest, the grain attracting the little rodent as well as the Chinese peasant. Rice growing results, of course, in an anthropic modification of the ecology, a modification that is involved in Korean hemorrhagic fever as well as in Japanese encephalitis, whose vector is also a mosquito. The latter disease strikes three hundred thousand people annually in Asia, leading to seven thousand deaths. This form of encephalitis is much more frequent during periods in which, in order to facilitate growth, the rice paddies are flooded and the mosquitoes proliferate. Argentinean hemorrhagic fever is due to the Junin virus, which emerged after pastures were replaced by cornfields, leading to the multiplication of the *Calomys* mouse. The number of cases among humans increased in proportion to the growth of the areas cultivated. With regard to influenza, we have already noted that Chinese farms, because of their peculiar biocenosis, promote the emergence of very aggressive viruses. The proximity of pigs, ducks, and chickens facilitates the exchange of genetic material among the influenza viruses of the various species, and thereby increases the danger that new forms of influenza will emerge. Arno Karlen, drawing on the work of biologists Thomas Hull and Evgeny Pavlovsky, reminds us that the list of infectious diseases due to domestic animals is very long. No less than sixty-five diseases are due to dogs, forty-five to cattle, forty-six associated with sheep and goats, forty-two with pigs, thirty-five with horses, and twenty-six with fowl. We should add thirty-two diseases associated with mice, which are not domestic animals but are nonetheless an undesired but constant companion of the human species. In all, we share about three hundred pathogenic germs with domestic animals and a hundred with birds and wild animals.

## ANTHROPIC ECOLOGICAL MODIFICATIONS ARE NOT THE ONLY SOCIOCULTURAL FACTORS IN THE EMERGENCE OF EPIDEMICS

Travel, migrations, and displacements of populations are, in the domain of human activities, essential factors in the emergence and/or diffusion of diseases. The term "globalization" has become commonplace because international exchanges are so frequent that the world has been transformed into a huge village in which the notions of quarantine and *cordon sanitaire* have become practically meaningless. To be sure, in long-ago periods such methods could be crudely applied and prove effective. But when confronted with large-scale movements of people, protective barriers became impossible to maintain, even in ancient times; at most, one could control entry through city walls or close access to a port. As we have seen in the preceding chapter, this did not prevent the advance of the Black Plague across Europe, and in Central America smallpox preceded the Spanish invaders in Mexico. Exchanges between Europe and Asia were continual and had various motives. The famous Silk Road is said to have facilitated the movement of rats and the plague.

The Crusades helped maintain the movement between Europe and the East. The slave-trade ships brought *Aedes aegypti* and yellow fever to the New World. Cholera, carried by travelers, left the Ganges for the rest of the world. Because of its geography and culture, Japan was an exception, the inhabitants of the Empire living ipso facto behind a permanent *cordon sanitaire*. Plague, which decimated the European population between the fourteenth and seventeenth centuries, spared Japan, where measles was also an almost totally unknown disease and one that was always imported. Today, Japan cannot, any more than any other country in our small world, prevent—technically, politically, economically, and especially definitively—travelers from invading its territory. The idea of frontiers hermetically sealed to "foreigners" strikes us as completely illusory in a world in which residents of Hong Kong can watch a live broadcast of the Monaco Grand Prix on their televisions and Icelanders can use the Internet to consult the menus of the best San Francisco restaurants. The international movements of people and goods have nothing in common with what they were in the examples just cited, unless it is that they can all promote the outbreak of epidemics. These outbreaks are all the more probable because the exchanges are so frequent and diverse and concern groups that are often very large. Human movements have played a major role in the dissemination of AIDS in Africa, and thus it has been written that "AIDS goes where people go." In central Africa, the main north-south routes and the railroads coming from Zaire and Tanzania meet in Zambia, the country of copper. In 1991, the rate of HIV-positivity reached 3 percent along the road to South Africa that passes through Zimbabwe; in 1993, in Harare, the capital, the rate reached 40 percent. To the east, the road passes through Malawi, where the rate of infection is the highest in Africa. Many of the country's inhabitants have worked or still work in South Africa's gold and diamond mines, and at least 50 percent of them were contaminated during their stay there or while traveling the road south to Johannesburg and Durban. This road is known as the "Road of Death," the "vector," if we can use that expression, being the prostitutes who work along it. For Joseph McCormick, an American physician working in Africa, the emergence of HIV took place in northwest Zaire, near Lake Victoria. To be able to determine objectively the virus's diffusion from that epicenter, we would have to have blood samples from soldiers, raped women, and prostitutes dating from the period of the conflicts between Tanzania and Uganda. The trucks moving from Dar es-Salaam and Mombasa toward Rwanda, Burundi, and Uganda crossed territories at war. At that time truckers and prostitutes played a role in the spread of the disease that we would have to be able to study.

Clearly, humans have a certain responsibility in the outbreak of new plagues. This is not a moral comment but only an incontestable observation, even if it seems peremptory.[54] By traveling, our species favors the dispersion of microorganisms. Among the pathogens whose dissemination is particularly connected with travel are the dengue virus, the filoviruses (Ebola and Marburg), the AIDS virus, and the yellow fever virus, as well as protozoa such as *Giardia* and *Plasmodium* (see following). Many of these germs are contacted through commercial exchanges; this was the case for the Marburg and simian Ebola viruses that resulted in the incidents in Europe and in the United States that we have already discussed. Sometimes transportation involves a vector that begins to transmit a local microbe when it arrives at a new site. In some twenty American states, we find Eastern equine encephalitis, which is due to a virus.[55] The virus can be transported by the *Aedes albopictus* mosquito, which flourishes on the American and African continents, where it was introduced from Asia along with imported used tires. The possibility that the cholera vibrio might be transported in ships' ballasts has already been mentioned in Chapter 1. It is thought that hundreds of species have been exchanged among different territories by this kind of transportation.[56] Let us not forget, in the context of this brief paragraph on trade, that mad cow disease was spread outside Britain not only because of the exportation of contaminated cattle but also because of the trade in animal feeds that were intended for cattle and that contained sheep prions. Human movements that can provoke outbreaks are not limited to international travelers, many of whom now travel by airplane. While it is true that every year there are five hundred million international travelers, we must not forget intranational movements, chiefly from the countryside to the cities, but sometimes in the reverse direction. It is estimated that by 2010, half the world's population will live in cities under conditions that are not necessarily satisfactory.

The end of the twentieth century was a period of "travels," migrations, and displacements of groups of people, sometimes on a global scale.

A study by Grubler and Nakicenovic shows that in two centuries (from 1800 to 2000) the number of kilometers traveled by the average French person increased a thousandfold. People often say that business and tourist travel represent only a small portion of human movements. However, as we said earlier, five hundred million humans annually cross their countries' borders on international flights. To these travelers, who range from tourists to missionaries, from businessmen to pilgrims, not to mention immigrants and students, we must add the great movements of populations such as those of legal and illegal workers, soldiers (on maneuvers or at war) and refugees. Regarding the latter, at the beginning of the1990s there were twenty million refugees and thirty million displaced persons worldwide. The consequences for the populations concerned can be extremely difficult and dramatic, as when in 1994 five hundred thousand to eight hundred thousand Rwandans took refuge in Zaire as a result of conflicts in their country, and fifty thousand died of cholera or dysentery caused by *Shigella dysenteriae*. Human migrations thus promote the emergence of infectious diseases through various mechanisms. Travelers can carry with them, or in them, microbes or vectors (e.g., lice).

In 1987, for example, Muslim pilgrims transported an agent of meningitis, *Neisseria meningitis,* from southern Asia to Mecca, in Saudi Arabia. At the holy site, other pilgrims picked up the microbe and carried it to Africa. In the areas where they settle, newcomers may be, because of their genetic predisposition or their customs, more susceptible to the local microorganisms while at the same time conveying germs to which they were resistant but to which the natives would hypothetically be very susceptible. Moreover, through their behavior, customs, beliefs, and technologies, the new arrivals could change the local ecology for the worse.

Let us remember as well that the means of transportation itself can be dangerous for those who use it. The most modern jets, in which the ambient air is, for economic reasons, slowly replaced and the hygienic quality of the services is not always very high, can spread Legionnaires' disease, cholera, intestinal infections, flu, and tuberculosis.

## DEMOGRAPHY, HUMAN BEHAVIOR, AND OUTBREAKS

The great movements of populations within countries or on a territory assimilated to a geographical or ethnic entity often follow constraints of human origin, whether these are physical threats (wars, revolutions, banditry) or economic pressures. Aside from situations of great political instability, as in countries undergoing revolutions or local conflicts, the most common situation is socioeconomic instability. The poverty and more than precarious ways of life of certain rural populations force them to settle around large cities, often under lamentable hygienic conditions, in the hope of finding simply the minimum that will allow them to survive. Soon the world's twenty largest cities will each have more than ten million inhabitants. Most of these cities are in "developing" countries, that is, places in which there is no way of ensuring sanitary conditions worthy of the name. This untrammeled urbanization is already clearly associated with outbreaks of dengue,[57] yellow fever, and malaria. We find two factors in all these outbreaks: demography and diseases due to vectors related to water. Intranational migrations seem to be of great importance in analyzing the demographic causes of outbreaks. The difficulty comes from the fact that we often have little detailed quantitative information about events of this kind. As we have seen, these movements are not only from rural to urban areas, but also sometimes in the opposite direction, for example in Brazil, where many city residents are moving to the Amazon area, which they see as an agricultural Eldorado. These new farmers, the poor among the poor, are burning the forests and exposing themselves to the local microorganisms, involuntarily generating a new pathocenosis. This reminds us that the rise of the Oropouche virus in Belém was caused by the unintentional creation of new habitats for the disease's vector. Those who have seen, in Manila or elsewhere, children spending long, difficult days trying to find something to eat by digging through veritable mountains of refuse, that is to say through immense incubators for diverse and various germs, will understand the microbial dangers of anarchical urbanization accompanied by extreme poverty.[58] In 1992, almost twenty million people worldwide were refugees, including a very large number of children wandering through the streets of cities, at the mercy of all the infamous maneuvers of adult criminals. Such a chaotic situation provides the ideal context for mosquitoes to reproduce—that is, one in which there are all sorts of containers filled with water (old tires, plastic bottles). The mosquito baby boom that arises from the disorder and insalubrity can only favor the diffusion of certain diseases, including, of course, those due to arboviruses (dengue, yellow fever, and so on).

Scientific and sociological reports, and even journalism referring to human factors in the emergence of infectious diseases usually associate demographic issues with some of the sexual behaviors for which the approaches that we can call scientific are still very inadequate. Sexual behavior is one of the human factors that is important, and sometimes very important, in the transmission of pathogenic microorganisms. Discussion of this problem is not easy, since it can be perceived as a vehicle for moral or even moralizing considerations that can paralyze any objective analysis. Without going into the matter further, let us say simply that human sexual behavior participates in the diffusion of pathogens, especially HIV, and it does so under three conditions; ignorance, coercion, and pleasure. In the preceding chapter, we mentioned the case of groups in Africa—but the same can be said of other places—that reject the reality of AIDS as a sexually transmissible disease. In such situations, which are fairly frequent in certain areas of the world, efforts at prevention seem to have little or no effect. The ignorance that underlies this or that behavior is associated with the most poverty-stricken societies. Coercion is frequently combined with ignorance and poverty. This coercion is social and economic in origin. Thai children who become prostitutes do not do so of their own free will; they are forced to do it because of their families' poverty and because of parental oppression. African prostitutes are no better off. With the notions of sexuality and coercion we must associate the situations observed during conflicts (Uganda, Zaire, Bosnia, for example) in which rapes are common and their consequences easy to imagine. We will not linger over the role of pleasure, but note only that the term may provide an all-too-convenient excuse for adults who engage in sexual tourism, thereby taking advantage of the poverty and helplessness of enslaved children. What would a well-off European or American father say if he found one of his children, whether a boy or a girl, in a brothel in Manila or elsewhere? Travel, and especially air travel, also plays a role in the diffusion of sexually transmissible diseases. In addition, urbanization can participate in the relationship between sexual activity and the emergence of infectious diseases. In many developing countries movement to urban zones, which is often motivated by economic concerns rather than a real desire to live there, is often accompanied by a rupture in the perception of customary values, a break with the parents' generation, and a rejection of the rules of traditional society.

We cannot mention human behavior associated with outbreaks of infectious disease without reference to drug use involving syringes and needles. Addiction to injectable substances is not in itself a factor in the transmission of pathogens; the danger arises when contaminated syringes are shared with infected addicts. Drug use is a calamity that we have not been able to overcome. Efforts to repress it seem ineffective when it is noted that in 1990 the number of people incarcerated in the United States for offenses related to drug use was higher than that of all federal prisoners in 1980. Here there is no point in quibbling over the moral aspects of drug addiction; we simply note that the latter can participate in the danger of infection.

## TECHNOLOGY AND "PROGRESS"

As we have seen, the outbreak of Legonnaires' disease illustrates a situation in which the pathogen took advantage of a cultural milieu produced by human technology, which was itself conceived in order to provide comfort, an artificial atmosphere, and a form of hedonistic pleasure. Examples of technology's contribution to outbreaks of disease are, of course, more numerous, and sometimes more pertinent than the example of Legionnaires' disease.

### *The food industry*

Among the technological advances likely to lead to outbreaks are those associated with biotechnologies involved in the mass production of foodstuffs. First of all, let us recall the history of mad cow disease. Farmers who fed their cattle sheep proteins were unaware that they were setting in motion a new process that would generate a variant form of Creutzfeld-Jakob disease. *Homo sapiens* is an omnivore: it has been a long time since most humans hunted and gathered in order to subsist. They now eat products that issue from processes of production, one of whose criteria, if it is not the chief criterion, is profitability. Modern industrial methods of producing foodstuffs are increasing in effectiveness, and decreasing in cost, but this entails the risk of contaminating consumers. It suffices that a microorganism is present in a sample of the raw material used in producing a foodstuff for the whole final product to be contaminated. Given the size, frequency, and increasing commonness of commercial trade, contamination can come from very far away and/or be distributed throughout the world. For example, the presence of the bacterium *Escherichia coli* in a few sides of beef has infected hamburgers in certain areas of the United States. The *E. coli* concerned belongs to the serotype 0157:H7, and provokes a hemolytic (red blood-cell damaging) syndrome, thus producing serious problems. It has become a thorn in the side of the American agro-food industry. *E. coli* is not the only agent responsible for infectious pathologies connected with food consumption.[59] The emergence of new pathogens has modified the epidemiology of infections originating in food. Many of these are present in animals that are in good health and provide the raw materials for food products (meat, eggs). It is thought that worldwide, these microbes cause millions of cases of sporadic illness or epidemics. Every year, millions of Americans fall sick with infectious diseases, and thousands of them die. Let us note that in order to feed humans adequately we have to regularly examine what the animals they consume are eating and drinking. Formerly, the foodstuffs involved in human infections were raw or insufficiently cooked products, whereas now contamination affects other products that are supposedly safe, such as eggs, which we now know can contain bacteria such as *Salmonella enteriditis* that are capable of surviving even when cooked in an omelet. In our time, alimentary infections no longer correspond to the cliché case of acute diarrhea following a first communion banquet or a friendly hunters' feast south of the Loire. The story of *Salmonella enteriditis* and the "egg connection" is exemplary of the modern aspect of outbreaks originating in food. The implication of *Salmonella enteriditis* was first suggested in sporadic cases in New England. In 1982, eggs were suspected of being involved in an outbreak in a retirement home. In 1986, this was confirmed after an epidemic affected thirty thousand persons in seven American states who had eaten egg noodles. Today, eggs are the main source of salmonella infections in the United States. We would be mistaken to ironize about such outbreaks. While they do not have the frightening, exotic aspect of an explosive disease emerging from an equatorial forest, they represent real dangers. Public health officials, even in a country with monitoring processes as elaborate as those of the United States, recognize that they have great difficulty in handling these epidemics. Moreover, the monitoring function will acquire a preponderant role in the way public health services manage epidemics large and small, whether of food origin or not. The old dream of eradication has given way to the more pragmatic procedure of monitoring (cf. Chapter 5).

### *Iatrogenesis*

Among the advances in technology that have just been mentioned, we must also note what is generally called "medical progress." In the next chapter, we will take up the crucial subject of the use of antibiotics and the emergence of resistant germs. This kind of selection arose from a form of medical practice dating from the discovery of penicillin, and which consists of fighting pathogens with medicines called antibiotics. We will emphasize that physicians are not the only dispensers of these molecules and that the emergence of resistant bacteria is a phenomenon that goes beyond the field of the medical use of antibiotics.

There are other domains of medicine in which technology has facilitated the transmission of infectious diseases. The one that has affected the most people, especially in France, is the tragedy of the transmission of the HIV virus through blood transfusions. Transfusion is a therapeutic act whose utility is no longer debated. One can be opposed to transfusion for religious or philosophical reasons, but hardly on the basis of medical considerations. Nonetheless, it is true that the use of human blood or its derivatives has favored the transmission of pathogens in those who have received it from a limited number of contaminated donors. Diffusion from a single donor to a large number of receivers resulted from the use of pools of blood products. Just as in the food industry a single egg carrying bacteria can contaminate a whole batch of pastries, the adjunction of contaminated plasma to a pool of blood taken from hundreds of different donors transforms a batch of human albumen or antihemophilic factor into products capable of transmitting an infection. Among the diseases that the practice of transfusion can transmit, we must cite AIDS; although systematic testing for HIV-positive donors has considerably diminished the danger, it has not been completely eliminated. Hepatitis C is a serious threat that remains persistent, whereas prevention of hepatitis B has benefited from systematic testing for contaminated blood samples, so that the risk of hepatitis B being transmitted by transfusion is virtually nil. Prions, as we saw earlier, might also be transmitted through the blood. If this is confirmed, another difficulty will be added to the range of diseases transmissible through blood, thereby further limiting the possibilities of the use of blood and its derivatives, for it is impossible to detect prions in the blood of donors who are apparently in good health but who will later develop Creutzfeld-Jakob disease. With blood transfusions we can associate organ transplants, which are also sometimes involved in the transmission of pathogens from the grafted tissues. The same was true of the use of growth hormones derived from the brain tissue of cadavers.

Under certain circumstances justified by the treatment of diseases, medical progress has caused the functioning of some patients' immune systems to be altered and their defense mechanisms against infectious agents compromised. The patients concerned are said to be immune-depressed, and they receive medicines that intentionally alter their immunological capacities. To this group of immune-depressed persons we can add, *mutatis mutandis,* patients with AIDS. Taken together, they represent a human group more susceptible to infections. The very elderly, who are growing more numerous in Europe and North America, also represent a group with diminished immunity, hence more susceptible to outbreaks. Finally, in many poor countries where there are refugees, food shortages are accompanied by problems with immunity that make children more susceptible to infections.

### THE EMERGENCE OF INFECTIOUS DISEASES IS FAVORED BY POVERTY AND SOCIAL INEQUALITIES

In May 1997, in one of the "Fact Sheets" it makes available on the Internet, the National Institute of Allergy and Infectious Diseases reported that since 1981, when AIDS appeared in the United States, minorities (African Americans and Hispanics) have constituted 54 percent of the 500,000 cases recorded. In 1996, men in these minority groups accounted for 56 percent of the male cases of AIDS recorded, while women in these same groups accounted for 78 percent of the female cases. In its July 1996 Fact Sheet, the NIAID had already noted that "Injection drug use (IDU) is a major factor in the spread of HIV in minority communities. During the period July 1994–June 1995, the rate of IDU-associated AIDS cases among African-American women was 31.8 cases per 100,000. The rate among Caucasian women was 1.9 per 100,000. The rate among African-American men was 78.7 vs. 5.8 for Caucasian men, 44.7 for Hispanic men, and 15 for Hispanic women. In that year, for every 100,000 African Americans, there were 89.7 cases of AIDS, or six times the number observed among whites (13.5 per 100,000)." This means that AIDS is the leading cause of death for African Americans of both sexes between twenty-five and forty-four years of age.

These data show the extent to which social inequalities, poverty, and as a result, ignorance, promote the emergence of infectious diseases. Such situations are not the monopoly of the poorest countries or so-called developing countries. In New York's Harlem, the mortality rate, relative to age, is higher in some groups than it is for the same groups in Bangladesh. In the next chapter, we will return to the relationships among poverty, social inequalities, and emerging diseases. For the moment, we will simply emphasize that this problem is not new. It was already discussed by René Dubos, a physician and humanist who recognized the importance of the social element in disease.

### CAN OUTBREAKS SOMETIMES BE CAUSED BY MICROBES ALONE?

We have discussed the role of human activities in the emergence of new diseases. Does that mean that all things considered, microbes play only a secondary role in outbreaks? Microbial evolution may favor the appearance of a more aggressive, more harmful agent. A mutation transforms a germ into a much more aggressive entity—that is probably what happened to *Yersinia pestis* before the great epidemic of the Black Death (cf. Chapter 1) and might explain other outbreaks as well. In the next chapter, we will examine the extremely important and worrisome problem of microbes that have become resistant to antibiotics. For the present, we should observe that there are few detailed examples in which the emergence of an infectious disease is strictly microbial in origin, that is, in which the prime mover was the rise of a pathogen with new, more virulent combative abilities. This possibility was suggested, but not formally proved, during the outbreak of Rift Valley fever. The flu virus might be considered the champion of continual mutation, yielding particularly dangerous forms almost every decade. Nevertheless, the mutagenesis of the flu virus is greatly facilitated by human behaviors. This is notably the case with animals raised together in Chinese poultry yards, which allows chickens, ducks, and pigs, amid piles of their excrement, to exchange their viruses (cf. Chapter 1). Isolated bacterial modifications are supposed to have been involved in the Brazilian purpuric fever caused in 1990s by a variant form of the bacterium *Hemophilus influenzae*. The same seems to be the case for the appearance of new, extremely dangerous forms of streptococci (cf. Chapter 3).

## CONCLUSIONS

From this brief review of certain outbreaks, their causes, and their consequences, it appears that they are complex, dynamic events in which germs paradoxically do not play a major role. Outbreaks are phenomena in which human activities have an essential place, whether in the domains of the economy, the society, technology, leisure, psychology, or even spirituality. We should, in fact, speak of new diseases rather than new germs, for the majority if not the totality of them are caused by already existing pathogens that have been thrust onto the scene of global infectious diseases, usually by humans themselves. These new diseases must be added to the infectious pathologies that are always present and to those that we thought had disappeared but are re-emerging in a disturbing way on the horizon of microbial threats present and to come.

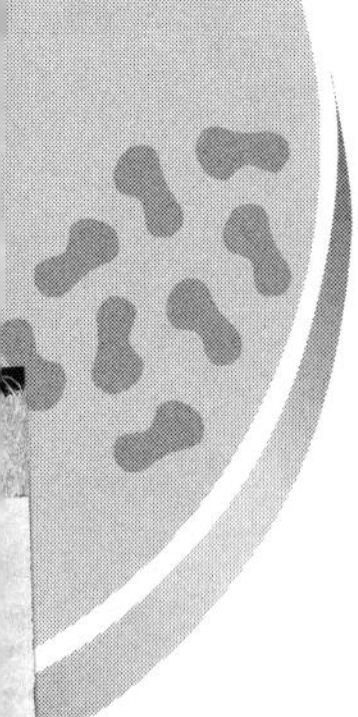

CHAPTER 3

# THE RETURN OF OLD DISEASES AND PERSISTENT DISEASES

## THE WHITE DEATH

In his very interesting autobiographical work, *The Hospital, a Patient's Biography* (*L'hôpital, une hostobiographie*), Alphonse Boudard tells the story of his tuberculosis and his long stay in sanatoriums and general hospitals. Apart from the fact that Boudard's book awakens many memories in those who experienced French hospitals up until the 1960s, it sums up the context of adult tuberculosis at the dawn of the use of antibiotic therapy to fight Koch's bacillus.[1] Everything in the hero's story is in agreement with what tuberculosis was, and how one contracted it: an adolescence spent under precarious material conditions during the German occupation, the unavoidable crowding during the campaign in Colonel Fabien's group, overcrowding and difficult living conditions again in prison. Then hospitalization, pneumothorax, injection of air into the pleura, surgery, and especially the administration of streptomycin and later on, isoniazid.[2]

Tuberculosis used to be called "the White Death." It is a very old disease that has recently been demonstrated to have been present in Egypt in the time of the pharaohs. It was passed from one person to another by droplets of saliva and pulmonary secretions, and the larger the group involved and the denser its population, the easier this transmission is. Tuberculosis was probably a disease that arose in the post-Neolithic age, since, in order to spread, it requires human concentrations of a reasonable size. In the middle of the sixteenth century, Fracastor was the first to emphasize the contagious nature of "phthisis" or "wasting disease," as tuberculosis was then called. The Koch bacillus strikes chiefly the lungs, but it can also affect any of the other organs; before the discovery of the

BCG (bacillus Calmette-Guérin) vaccine, tubercular meningitis was a major cause of death among children.

Over the past two centuries, the White Death has killed some two billion humans. The bacillus infects one person out of every three worldwide, or about two billion *Homo sapiens*, resulting in about two million deaths per year. In short, despite antibiotics the Koch bacillus annually kills one person out of every thousand infected, and one person with the disease in ten. It is easy to imagine how relieved patients and doctors were when the first antibiotics (including streptomycin and isoniazid) came into use. These miraculous medicines allowed the closure of "TB" sanatoriums, the first of which had been opened in 1850 by Dr. Hermann Brehmer in Görbersdorf (then in Germany, now in Poland). The creation of sanatoriums shows to what extent tuberculosis was a disease that affected a large number of people and that society felt called upon to combat. Hospitals for tuberculosis patients varied in the degree of comfort they provided, ranging from those that met basic needs to the kind of medicalized hotel described in Thomas Mann's *Magic Mountain*.

According to the WHO, tuberculosis is currently the chief killer of adults in the world; in 2003, 1.7 million people died of the disease. Thus, it is said that tuberculosis is "making a comeback": a disease we thought had been eradicated, or at least was about to be eradicated, is present on every epidemiological battle line. In France, the battle against tuberculosis has a long history; it now belongs to France's cultural heritage of the history of medicine and the struggle with disease. "Historical" tuberculosis has one thing in common with the illness we currently see, namely, the fact that *Mycobacterium tuberculosis* is a bacillus that strikes the weak. Just as the hyena in the savanna attacks only feeble animals, the Koch bacillus preys on humans whose immune systems have been weakened (as is often the case with persons who have AIDS), whose general state of health is impaired (those suffering from malnutrition and other deficiencies), or who suffer from a lack of affection or from emotional shocks (tuberculosis used to be considered an illness related to unhappy love affairs). The current upsurge of the disease has various causes. In the United States, there are thought to be at least four main causes: HIV infection, the immigration of carriers from countries where the disease is common, transmission of the disease in overcrowded situations (hospitals, prisons, shelters for the homeless), and the deterioration of public health monitoring systems. According to American health authorities, the preceding observations are associated with a significant abandonment of antituberculosis programs. In 1989, the Centers for Disease Control and Prevention (CDC) was predicting that tuberculosis would be eliminated in the United States by 2010. As early as 1992, however, there was general agreement that things were not developing as expected, because the number of cases was increasing and more and more cases of antibiotic-resistant tuberculosis were being seen. The emergence of a new form of the disease led authorities to elaborate a plan, not to eradicate the infection, but simply to combat the new tuberculosis.

It is easy to understand the dismay of authorities at the CDC. Since the end of World War II—that is, during a period in which antitubercular drugs were regularly used—the number of cases had steadily diminished at the rate of 5.6 percent per annum, from more than 84,000 cases in 1953 to 22,255 in 1984. However, from 1985 to 1993, the incidence of the disease increased by 14 percent, from 22,201 cases to 25,313 cases. Many carriers of the tuberculosis bacillus show no symptoms of the disease, and it has been estimated that from ten million to fifteen million Americans are infected with *Myobacterium tuberculosis.* In fact, as early as the beginning of the 1950s, professional journals addressed to members of the health professions such as general practitioners were alarming their readers by suggesting that tuberculosis might be coming back. This information has now reached the general public, and officials are organized to fend off the attack.

The "new" Koch bacillus is drug-resistant or, more often, multi-drug-resistant (MDR).[3] Resistant bacteria, whether the tuberculosis bacterium or others, have the ability to transmit their know-how to bacteria that are not yet resistant but that will become resistant after they have inherited MDR capacity. Resistant bacteria communicate MDR information to their neighbors in the form of a short segment of DNA that carries the message. This DNA can travel in the form of plasmids, small bits of circular DNA, harmful "Frisbees" that are tossed back and forth between microbes (a process known as "conjugation") or in the form of transposons, small genetic elements that can "jump" from one chromosome to another.[4] Sometimes viruses that infect bacteria can transport genetic information from one bacterium to another. As Courvalin has written, "We must always keep in mind that the opportunities for the exchange of genetic material among organisms in nature are immense."

Most cases of resistant tuberculosis are caused by bacteria that can resist isoniazid and rifampin, which are usually the effective medicines. Among the first victims of this kind of tuberculosis, many were HIV-positive and had been treated with strong doses of medication. In hospitals, resistant forms of the tuberculosis virus have been transmitted to the staff. This kind of well-known interpersonal communication of the disease explains why among prisoners the average number of cases is four times higher than among other groups. In prisons contagion is promoted by extreme overcrowding, poor air circulation, the high frequency of HIV-positive prisoners, and the size of the contaminated population. Poor social conditions once again favor contamination; in the United States, 84 percent of cases of tuberculosis among children occur among disadvantaged ethnic groups.

Overpopulation, lack of hygiene, and insalubrious housing have always promoted tuberculosis. In the nineteenth century, urbanization under conditions that were often deplorable encouraged tuberculosis, which was then one of the most common causes of death in Europe and North America. Overcrowding in enclosed spaces with poor air circulation is such an important factor in disseminating the disease that there has been contamination among airline flight personnel (a flight attendant with tuberculosis transmitted the Koch bacillus to twenty-three of her colleagues in the course of her career); transmission among passengers has also been reported.

The reader will have noted that some people are contaminated by the tuberculosis bacillus without falling ill, whereas others show symptoms of the disease.[5] To contract tuberculosis, one must, of course, be in contact with the bacillus, but one must also have, for one reason or another, a susceptibility to the disease. We have already mentioned the susceptibility connected with social origin, and we will return to the possibility of a biological susceptibility to tuberculosis. Here we can note that some people have suggested that ethnic predisposition plays a role in the occurrence of the disease. A more scientific approach consists of studying the genetic determinism of response to the antigens of the Koch bacillus. No one has yet clearly demonstrated the existence of a biological trait associated with either susceptibility or resistance to tuberculosis.

Vaccination with BCG, widely practiced in France, has not been applied everywhere in the world—far from it. In the United States, for example, BCG is not advised by the CDC except in the case of children who are not allergic to tuberculin and who are in regular contact with tubercular individuals. In many countries the utility of this vaccine is a matter of debate. BCG is surely not the ideal vaccine, and many laboratories are working to develop a new one. But here we must not throw out the baby with the bathwater; let us recall that BCG has certainly prevented the occurrence of serious tubercular primary infections.

At present, the danger proceeds from resistant forms of tuberculosis, to the point that when the 150th anniversary of Robert Koch's birth was celebrated in Berlin in 1993, the facts we have noted and pessimistic predictions hardly gave health authorities attending the celebration reason to be proud. The battle against tuberculosis is thus far from being over. A reminder issued by the WHO (EURO/01/05) noted that after the African region, the European region has the lowest rate of successful treatment of individuals recently infected by the tuberculosis bacillus. In the European region, the chief result of tuberculosis is "polypharmacoresistance" (the agency's term for multiple drug resistance). Indeed, the unregulated and unmonitored use of antituberculosis drugs generates forms of resistance; when all the consequences are taken into account, it is less harmful to the patient and to the community to leave a case of tuberculosis untreated than to treat it incorrectly. To deal with this problem, the WHO has come up with an international TB control strategy known as DOTS (Directly Observed Therapy—Short Course). DOTS calls for sustained political commitment, access to quality-assured sputum microscopy, suitable chemotherapy, and an uninterrupted supply of drugs, as well as a recording and reporting system that makes it possible to evaluate the effects of treatment.[6] By 1998, 119 countries, including 22 recognized to be at high risk, had adopted this strategy. The results have been uneven. In countries with favorable results, from Peru to Vietnam, the therapeutic success rates are more than 85 percent, and detection of carriers of the disease is as high as 70 percent. Since 1991, when the DOTS strategy was formulated, more than thirteen million patients have benefited from treatment. There is no doubt that this strategy has greatly improved the situation: between 1994 and 2003, more than seventeen million tuberculosis patients were treated using the DOTS protocol. The therapeutic success rate was 82 percent in two countries where tuberculosis poses a major threat, so that we can reasonably hope that in 2015 the incidence and prevalence of tuberculosis, as well as number of deaths from the disease, will have decreased.

## LEPROSY

### *History*

Leprosy, also called Hansen's disease,[7] is a chronic bacterial illness affecting the skin, the mucous membranes in the upper respiratory system, and the peripheral nerves. Hansen's bacillus (*Mycobacterium leprae*), which is responsible for the disease, is related to the tuberculosis bacillus; they are both members of the Mycobacteriaceae family. The disease can take two extreme forms, lepromatous leprosy and tubercular leprosy, along with intermediary forms. Leprosy is not a disease that affects Europeans, but it remains extremely common in some parts of the world. India, Indonesia, and Myanmar (Burma) have 70 percent of the worldwide total, which is estimated at between ten million and twenty million, including about six hundred thousand new cases per year. In twenty-eight countries, the prevalence is greater than 1 percent, so that 1.3 billion humans are exposed to the disease (though not all of them fall sick with it). In epidemic areas, many people are certainly exposed, but fortunately 95 percent of them eliminate Hansen's bacillus without contracting the disease. In short, *Homo sapiens* resists leprosy fairly well, and it has been demonstrated that the majority of those who get sick have a constitutional predisposition related to a deficient immune response to the bacillus. Although it seems incontestable that some people are predisposed to contract the disease, it remains that leprosy is now chiefly a problem of poor countries, a pathology of the word's "poverty belt," whereas it used to affect every continent. Even after its disappearance in some countries, leprosy has left a terrible mark on the collective memory; in many ways it is still a disease whose public image was and remains unique. In a time when we are often concerned with those who are excluded from society, let us emphasize that in Europe, and especially in the Europe of the Middle Ages, lepers were an excluded group. In the Bible, leprosy already is associated with the supernatural, with uncleanness and sin; from the Pentateuch to the New Testament, leprosy is frequently mentioned.[8]

In practice, lepers lived apart in leper houses set up outside the cities. They were allowed to move about but had to warn people of their approach by means of a clapper or whistle. They were in a sense exiles, taken away from their families after a medical and judicial examination. To complete their separation from public life, a mass was said for them that resembled a funeral mass.[9] Lepers were treated in such an unenviable way not only because it was feared, understandably enough, that they would spread the disease but also because of their supernatural and allegorical aspect. The leper could also be considered a member of the elect. After all, for Christians, the leper Lazarus was loved by Christ, and the dereliction experienced by the leper might serve to redeem other humans.

Since ancient times, it has been understood that leprosy could be transmitted directly from one person to another, and in fact the only possible reservoirs are the armadillo and certain apes, which were unlikely to contaminate *Homo sapiens* in Europe during the Middle Ages. The danger comes from the fact that leprosy develops slowly. The incubation period is long, and a person who is not yet sick can be contagious and transmit Hansen's bacillus. The bacillus is consequently very well adapted to the human species. Torpid, it quietly establishes itself and manifests itself very little if at all, using its new host only as an innocent vector. In Europe, leprosy decreased greatly after the Black Death epidemic. The generally accepted view is that the demographic collapse caused by the plague hindered interhuman transmission of leprosy. Another possibility is that some form of susceptibility or resistance crossed between leprosy and plague. On this hypothesis, those who were susceptible to one disease were also susceptible to the other one and were eliminated; or, inversely, the only survivors were those whose immune systems responded to plague (and leprosy).

In modern times, leprosy has long been treated with an antibiotic, dapsone, which is effective and inexpensive. Unfortunately, in 1977 a *Mycobacterium leprae* resistant to dapsone was reported. This initial observation was later confirmed, and resistant strains of the leprosy bacillus are being found with increasing frequency. The alternative is to use another medication, rifamycin, but strains resistant to it too have already been found (a multiresistant strain of *M. leprae* was discovered in Ethiopia). In Africa, Hansen's bacillus flourishes on a continent where the few resources available for public health concerns are spent on AIDS, malaria, and other tropical diseases. The situation is also worrisome in Brazil and Colombia, and there are sporadic cases in central and eastern Europe as well. With the emergence of multiresistant bacteria, leprosy is once again a genuine public health problem throughout the world.[10]

### *Susceptibility of biological origin*

Does leprosy rest on a biological substratum explaining why it is regularly observed that most humans are immunized against it and that only a minority fall sick with it? In fact, those who are sick present a special kind of immune response to the disease. Macrophages (phagocytic cells that can engulf and destroy bacteria, viruses, and other foreign substances; see Chapter 4) in patients suffering from lepromatous leprosy are unable to destroy *Mycobacterium leprae,* which, like *Mycobacterium tuberculosis,* sets up shop inside them. These patients' immune response is not strong. Immunity to lepromatous leprosy is connected with traits in the HLA (human leukocyte antigen) system. Patients having the HLA-DR3 trait are more likely to contract tuberculoid leprosy. In addition, in these patients' families, those who fall sick are often the ones who have the same traits.[11]

Although this confirms that the patients are persons whose genetic programming prevents them from responding effectively to leprosy's antigens, it is legitimate to question the pertinence of the protection that might be provided by a vaccine. Would someone be immunized by a vaccine if he cannot immunize himself against the pathogen itself?

## MALARIA IS BACK

Like yellow fever, malaria was one of the serious dangers that confronted Europeans tempted by colonial adventures. France, Britain, and Portugal adopted colonial policies that exposed their armies and their colonists to a dreadful parasitical disease that proved very deleterious for the European invaders. It was a military doctor, Charles Louis Alphonse Laveran, who, while accompanying French troops in Algeria in 1880, discovered the parasite responsible for the disease in the blood of twenty-six soldiers out of forty-four who had fallen victim to malaria. Malaria is in fact a parasitical disease that is currently considered eradicated in North America and Europe, but in the nineteenth century it was well known throughout the Mediterranean basin. Moreover, in 1881 Italian physicians made a decisive contribution to Laveran's work, for in Italy there were many patients suffering from "fevers."

As *Time* magazine indicated in 1993, malaria is back. This return is associated with the parasite's resistance to drugs, and the hope of eradicating this parasitosis now seems completely utopian. No less than four species of *Plasmodium* infect humans: *P. vivax*, *P. ovale*, *P. malariae*, and *P. falciparum*. The latter is the most aggressive and most often involved in deaths resulting from malaria. Malaria is a major problem for public health worldwide.[12]

According to the WHO's latest estimates, in 1997 three hundred to five hundred million persons worldwide were infected with malaria. This parasitosis is reported to have caused 805,300 deaths in sub-Saharan Africa in 1990. Currently, malaria results in one million to two million deaths a year. Formerly very widespread, malaria is now confined to the poorest regions of Africa, Asia, and Latin America. It has been around since time immemorial; the utility of quinine in treating it was known before the etiology of the disease was discovered. Although Sir Patrick Manson mentioned in 1894[13] that malaria was transmitted by mosquitoes, in the fifth century B.C. Herodotus had already noticed that in the swampy areas of Egypt people slept under mosquito nets or at the top of "towers" where they were protected from mosquitoes. All peoples were not so well-advised, and malaria was most often attributed to "miasmas" that corrupted the air near swamps. For more than twenty centuries, the Chinese have treated malaria by drinking a tea made with qinghao (*Artemisia annua*), but the active ingredient was not identified until recently; it is called artemisinin.

The bark of cinchona tree (*Cinchona ledgeriana*) was used in pre-Columbian South America to reduce fevers. Its first known use to treat malaria dates from 1600, in Peru.[14] It was later used in England, where it was called "Jesuits' powder" or "Jesuits' bark." Cinchona bark was effective against malaria because it contains quinine. In 1820, Pelletier and Caventou succeeded in extracting quinine from the bark.[15] Quinine quickly became a drug used throughout the world to prevent malaria and perhaps to relieve its symptoms. In 1942, the invasion of Java by Japanese troops deprived the Allies of quinine. They substituted chloroquine, which is still more effective. Currently, because of *Plasmodium*'s resistance, chloroquine is increasingly being replaced by mefloquine. Resistance to chloroquine first appeared in Asia in the 1950s. Then doctors turned to mepacrine, proganil, and pyrimethamine. In the 1960s and 1970s, after the Vietnam War and the regular use of these drugs, most of them had almost ceased to be effective against malaria. The mefloquine now prescribed has led to new forms of resistance. In Thailand, it was observed as early as 1991 that a *Plasmodium falciparum* was resistant to halofantrine, a new antimalarial molecule that had never been used in the country. That is the most fascinating thing about it: the protozoon can resist new drugs because it is equipped with membranal pumps that expel every drug that enters its organism. *Plasmodium* is now more and more often resistant to multiple drugs.

To sum up the modern history of malaria, we must go back to the 1950s, the glorious age of the battle against infectious diseases. In 1955, the WHO launched the Global Malaria Eradication Program, which included using insecticides inside houses. The results were favorable in North America, southern Europe, the former USSR, and parts of Africa and Asia, where the disease was apparently eradicated. Unfortunately, in Asia and Latin America, where the results were less definitive, it was soon seen that malaria persisted; it had to be admitted that the project of eradicating malaria in Africa was illusory. In Sri Lanka, where malaria had been thought to be wiped out, people's enthusiasm was cooled by a major epidemic of the disease that occurred in 1968. There were new outbreaks in Central America and Asia, and in 1988, more than twenty-five thousand people died of malaria in Madagascar. By 1969, the hope of eradicating malaria was generally abandoned.

At the present time, malaria is endemic in ninety-one countries, but 80 percent of the cases are recorded in tropical Africa, where *Plasmodium falciparum* is dominant. We must recall that malaria often kills children: eight hundred thousand die of it every year. The socioeconomic consequences are catastrophic, especially in poor countries. In tropical Africa, for instance, the total cost of malaria (prevention, care, economic impact, etc.) is estimated at 1.8 billion dollars.

For many African countries, we can review the litany of causes and consequences of malaria. Epidemics are accompanied by social and economic repercussions, especially in rural areas. Many population movements are from areas infested with malaria toward others where malaria is less prevalent. These movements are motivated by economics (seasonal workers), demographics (farmers), politics (refugees), and so forth. Often the new arrivals do not have immunity to local diseases and will contract a serious illness; the situation will be all the more worrisome because in these countries the hygienic measures taken are rarely up to dealing with the needs.

Are things really so bad? If the project of worldwide eradication seems utopian, we should still note that in 1955 malaria was endemic in 150 countries, sixty more than is the case today. Of the five and a half billion inhabitants of our planet, almost 60 percent need not fear contamination. Thirty-three percent live in regions where endemic malaria has considerably lessened. Nine percent of humanity, some five hundred million people, the majority of whom live in Africa, are still seriously threatened by *Plasmodium falciparum*. The danger is, of course, individual, especially for children, but as we have emphasized, it is also biological and socioeconomic in proportion to the increasingly large number of serious forms of malaria connected with the parasite's resistance to drugs. Humans are responsible for many African epidemics resulting from wars, revolutions, population displacements, and deterioration of the natural environment.

### *Biology*

Malaria is a sickness that manifests itself in periods of fever and chills alternating with periods of remission. Its beginning is often insidious, and headaches, a general malaise, and a slight fever make people think they have the flu. The periods of crisis are extremely painful and debilitating, but normally not life-threatening. It is the grave forms of the illness, the cerebral forms that are accompanied by a coma and very high fever, that can be fatal.

The biological cycle of the protozoon plasmodium allows us to understand the progress of the disease. The parasite's sex life takes place in the female *Anopheles* mosquito, and its asexual (or schizogonic) life takes place in humans. In *Homo sapiens* the plasmodium undergoes an erythrocytic stage (meaning in the host's red blood cells) and also a hepatic stage (in the host's liver). Thus plasmodium's biological cycle is fairly complicated, and it involves the female *Anopheles* mosquito, which is the only mosquito that needs the blood of mammals to nourish itself. While it is eating, the mosquito injects the plasmodium, along with its contaminated saliva, into the human. The plasmodium settles in the host's liver and develops there. The infected liver becomes a deposit of merozoites that are then let loose in the circulating blood, where they infect red blood cells. Within these cells, the merozoitic plasmodium is transformed into a schizocyte and then into a gametocyte that is absorbed by another *Anopheles* mosquito. In the female mosquito, fertilization produces zygotes from which proceed sporozoites ready to contaminate other humans, and so forth. The mosquito elicits an immune response in humans that can, after a long time, lead to a regression of the parasite. Moreover, the protozoon needs the hemoglobin in human red blood cells in order to feed itself, and these cells must be "of good quality." This is the case for individuals with abnormal hemoglobin[16] or an enzymatic deficiency such as a deficit in glucose-6-phosphate dehydrogenase. Sometimes the parasite cannot penetrate the red cell's membrane. Here we are concerned with biological circumstances that prevent or slow infection by the parasite. Therefore, this is not a matter of "conventional" immunology. Apart from these problems of nonimmunological human biology, it is clear that the mosquito plays a crucial role in human malarial infestation. In order for the plasmodium to reproduce, the female *Anopheles* mosquito has to live for twelve days, a not negligible length of time if we consider the mosquito's normal life cycle. It has been shown that prolonged survival of the vector can increase by a factor of fifty the number of cases of malaria it causes. This kind of situation is seen in equatorial Africa and southeast Asia where, because they live longer, it takes only a few mosquitoes to transmit the parasitosis, and thus a small number of parasites can constitute an almost inexhaustible reservoir of plasmodium. In view of these facts, fighting malaria by eliminating the *Anopheles* mosquito requires a powerful and effective insecticide. Unfortunately, pesticide-resistant mosquitoes are now being found with increasing frequency, to the point that some teams are trying to replace the mosquito that harbors the parasite with another mosquito that has been genetically modified and is resistant to plasmodium. Plasmodium's biological cycle includes a necessary stage in the mosquito, which amounts to a genuine biological interaction between the two species, the mosquito being not just a "flying syringe." Any way of introducing mosquitoes resistant to the parasite would be helpful in fighting the disease. It has been recognized that to combat malaria effectively, we have to have several irons in the fire and must not fail to take classical hygienic measures such as precautions intended to prevent mosquito larvae from proliferating in bodies of water. Such measures have enabled the Chinese to limit the invasion of malaria.

On the level of human ecology, the long cohabitation of the parasite and humans has led to phenomena of selection connected with five factors: hemoglobin, the constitution of glucose-6-phosphate dehydrogenase, a blood group known as "Duffy," a set of traits involved in the rejection of human grafts (the HLA system), and finally a genetic system peculiarly susceptible to malaria. Haldane was the first to note associations between a biological trait and the parasite. His observations were remarkable—and in agreement with the neo-Darwinian model of adaptation in the genetics of populations. The strength of the association between the nature of hemoglobins and resistance to malaria is such that it serves as a descriptive model of polymorphism that is stable, remaining unchanged over time.[17] We owe to Haldane the now-confirmed hypothesis that heterozygotes for S-hemoglobin resist malaria.[18] In Africa, the zones where plasmodium is endemic coincide more or less with those where sickle-cell anemia occurs. Malaria more often strikes individuals with normal hemoglobin than those who have the hemoglobin connected with sickle-cell anemia. A similar phenomenon is seen in other abnormal hemoglobins, as when red cells are deficient in glucose-6-phosphate dehydrogenase. Another genetic marker associated with resistance to malaria is that of the Duffy blood group.[19] It has been proven that molecules bearing Duffy traits, carried by the red blood cells, serve as receptors for certain plasmodia (*Plasmodium vivax* and *P. knowlesi*).[20] Consequently, being Duffy negative means that one does not have the "entry point" for *P. vivax* and *P. knowlesi,* and is protected against it. Moreover, in Africa, in territories where Duffy-negative populations live, *P. vivax* is absent. It is hard to say whether the absence of the Duffy trait has prevented *P. vivax* from establishing itself or the parasite has put selective pressure on Duffy-positive individuals, meaning that individuals with the Duffy trait have been eliminated. Thus genetics and evolution confer on individuals biological traits that make them more or less resistant to invasion by the malarial protozoon. Other genetic traits that create predispositions to malaria or tend to prevent malaria probably exist. There is said to be a gene that predisposes individuals to infection by plasmodium, as was described in an investigation carried out in Cameroon. Similarly, relationships between malaria and the HLA tissue traits involved in the rejection of grafts have been reported.[21] It has been shown that defenses against malaria are correlated with HLA traits. For example, an HLA trait called HLA B53 diminishes by 40 percent the risk of contracting malaria, especially a serious form of malaria; moreover, in Africa the B53 trait is more frequent among normal individuals than among those suffering from a serious form of malaria. This is probably why the trait in question, which is associated with a decisive selective advantage (resistance to malaria) is more frequent in sub-Saharan Africa (40 percent of Nigerians have the B53 trait) than in Europe (only 1 percent of French people carry the B53 trait). While there is no question that, by nature, HLA molecules can optimize defenses against malaria, it has been shown that various strains of *Plasmodium falciparum* that are theoretically susceptible to the immune system's cells can cooperate and ensure that only one of the two is eliminated, the other persisting in spite of the ad hoc HLA molecules.

Genetically determined protection against malaria may combine several of the elements mentioned above, such as hemoglobins, the enzyme glucose-6-phosphate dehydrogenase, and HLA molecules. The whole set of defenses can bear the stamp of neo-Darwinian evolution and, in short, represent an almost academic model. On the plains of Sardinia, for instance, we find people having HLA traits associated with resistance to malaria, and the frequency of these individuals who have fetal hemoglobin or who have a deficit in glucose-6-phosphate dehydrogenase varies in relation to the earlier malarial infestation of this part of Italy.

### *Malaria and human societies*

Not everything in outbreaks of malaria, as in other epidemics, is biological in nature—far from it. Geographical, socioeconomic, and cultural factors are very important. For example, an investigation of 445 Sudanese children who were younger than five years old and had been exposed to malaria suggested that malaria was associated with malnutrition, overcrowded living conditions, and low levels of education among parents, and was inversely related to the number of rooms in the home and to the possession of a computer. There are situations in which cultural circumstances alone can affect susceptibility to malaria.[22]

In the case of malaria, it is possible to find the same ecological, anthropic, sociological, cultural, and other causes that we have emphasized in the emergence of new diseases. For example: in Central America the development of agriculture, irrigation, and resistance to insecticides has led to a new increase in cases of malaria. In Brazil there are sixty thousand cases of malaria per year, or more than half the cases of this disease reported in the Americas. Malaria causes between six thousand and ten thousand deaths a year, most of them linked with the establishment of new farms and mines in the Amazon region, which leads to the creation of many new reservoirs of water propitious for the growth of mosquitoes. In Khartoum, fifty thousand cases of malaria occurred after unusually heavy rains and flooding. In Africa, as in Asia, cases of resistance to chloroquine are increasingly frequent, and in the savanna and the forests more than half the population is contaminated; malaria is the chief cause of infant mortality, killing 5 percent of children under five. In 1998, in Madagascar, twenty-five thousand deaths resulted from an epidemic connected with changes in agricultural techniques and climatic disturbances. In Afghanistan, there are three hundred thousand cases per year because of the total abandonment of public health measures and movements of populations connected with the civil war in the 1990s. In Cambodia, Laos, Burma, Thailand, and Vietnam, where the problem of forms of malaria resistant to drugs is the most acute in the world, the risk of epidemics is constantly increasing.[23] In the Philippines, Vanuatu, the Solomon Islands, and Papua New Guinea, three hundred thousand cases a year are reported, and these are essentially related to the occupation of new territories.

### *Anti-malarial measures*

Very early on, efforts were made to produce an antimalaria vaccine. Unfortunately, the production of a vaccine against malaria, like all vaccines against diseases caused by parasites, is extremely difficult, not to say impossible. A Colombian physician, Manuel Patarroyo, has proposed using a vaccine consisting of a protein with antigens already existing in various forms of the parasite. This vaccine, which is still in the experimental stage, is of debatable efficacity. Thus malaria, a very ancient disease, escapes therapeutic means and circumvents the cures invented by humans. Like many tropical diseases caused by parasites, malaria does not produce an effective immunity in humans, one that is curative and protective. At most, the immunity it produces can prevent the paraiste's proliferation from becoming too extensive, because each immune person represents an obstacle to the spread of the disease. Since the *Anopheles* mosquito can resist insecticides and since humans constantly promote the proliferation of the harmful partners plasmodium and the *Anopheles* mosquito, there is every reason to think that the triad of humans-mosquitoes-plasmodium, which Braudel calls the "malaria complex," will persist for a long time as an element inimical to human well-being.

The emergence of malaria in new areas is essentially anthropic in origin, and it is related to the conquest of new territories by means of deforestation. In Chapter 5, we will discuss a new way of conceiving of our relationships with infectious diseases. We will have to give up the utopian hope of eradication in order to make room for the pragmatism of forecasting. Technologically sophisticated methods employed on a global scale should allow us to predict, for example, an increase in the spatial and temporal distribution of a vector.

## ANTIBIOTICS: HOPES AND REALITIES

In 1942, Anne Miller was admitted to New Haven Hospital (Connecticut) for a strep infection contracted from her son, who had a strep throat.[24] The patient presented septicemia[25] with pelvic metastases of the infection. Her fever had risen to 108° F, and her doctors were desperately trying the treatments available at the time, including blood transfusions and even extracts of rattlesnake serum. On March 14, the physicians on duty received a bottle of penicillin from England. At 3:00 P.M. they began a revolutionary new treatment, using intramuscular injection of the antibiotic every eight hours. The following day, at 9:00 A.M., after she had received 36,000 units of penicillin, not only did Mrs. Miller have no fever, but all her vital signs were normal. That afternoon she sat up in bed . . . In 1996, Mrs. Miller was eighty-five years old.

Today, many people will find it difficult to imagine the excitement such an event aroused in the 1940s. A miraculous drug had been discovered. Physicians finally had at their disposal a therapy that would enable them to overcome infections, something that had been sought like the Holy Grail of infectology.

In 1942, pneumonia could be cured with ten thousand units of penicillin. The bacillus involved in Mrs. Miller's case, a Group A streptococcus, also caused scarlet fever. But under the effect of antibiotics, it had practically disappeared by the 1960s.[26] Unfortunately, the bacillus that had "disappeared" was replaced by its cousin, Group B streptococcus, which was more virulent and more dangerous for infants. Moreover, during the 1980s, a new form of Group A streptococcus appeared that was resistant to conventional antibiotics, while its cousin in Group B remained a danger for newborns. In 1992, it took on average twenty-five million units of penicillin a day to treat a streptococcus infection, and even then there was no guarantee of a cure. Group A streptococcus has become so threatening and can produce such lesions in tissues that in English-speaking countries it is designated by the acronym IGAS (Intensive Group A Streptococcus) but is also called the "flesh-eating bacterium." In the spring of 1990, Jim Henson, who created the Muppets, died as a result of an infection with a streptococcus resistant to antibiotics. The bacterium in question produces an extremely dangerous toxin through a mechanism known as a superantigen.

Another bacterium was to follow, so to speak, a similar path of resistance: pneumococcus.[27] When it was first used, penicillin allowed physicians to cure any pneumonia caused by pneumococcus—easily, meeting no resistance. By the 1980s, however, it was clear that certain strains of pneumococcus were resistant to penicillin and were particularly involved in ear infections. In 1982, there was an epidemic of pneumonia in Salt Lake City that lasted from 1982 to 1985, with a rapidly growing number of young victims, despite the massive use of antibiotics. In Oklahoma, in the 1990s, there was an epidemic of pneumococcal pneumonia resistant to antibiotics. Once again, the disease's victims lived in close quarters, such as elderly patients in hospitals and young blacks in disadvantaged areas. Pneumococcus also produced a new complication: a form of rheumatism that affects children and adolescents. This form of rheumatism was to spread throughout the United States.

According to the WHO, in 1992 about two billion children suffered from respiratory infections; among these, 4.3 million died from their illness, 800,000 of them from pneumococcal pneumonia.[28] Among the pneumococci, some are particularly worrisome, such as type 19A, which appeared in Durban, South Africa, in May 1977. At the very beginning among hospitalized children, three out of five died. The pneumococcus involved was resistant to the fifteen major antibiotics available at the time and sensitive to only one: rifampin. This one sensitivity, which ultimately permitted the outbreak to be controlled, did not prevent the bacillus from spreading to other pediatric clinics in Durban and Johannesburg, infecting children and staff. We see that pneumococcus, like many infectious agents, can be transmitted in a hospital environment.

Infections contracted in hospitals, called "nosocomial" infections, are one of the banes of modern health care.[29] *Staphylococcus* is among the bacteria that bedevil health care personnel and endanger hospitalized patients. In the United States two million patients per year contract nosocomial infections, and these lead to sixty thousand to eight thousand deaths annually. Thirteen percent of these infections are due to *Staphylococcus*.[30] The bacterium is transmitted by skin contact with the hospital staff, but it is also found on clothing. Although of course clothing, as a "second skin," plays a protective role as well, it nonetheless provides an excellent habitat for bacteria and for parasites that live on the skin.

Staphylococcus secretes toxins the effects of which can be extremely harmful, particularly because some of them are superantigens. Superantigens, which overstimulate the human immune system, provoke a real explosion in the population of the cells in our immune system by unleashing violent substances called cytokines that flood the organism and produce what physicians call toxic shock syndrome. This disorder was first described in 1978. It was found among menstruating women, a significant percentage of whom died quite rapidly from a disease that struck them like a thunderbolt in the middle of lives that had up to that point been normal. It was noted that many of these women had been using superabsorbent tampons. These tampons, which could be worn longer than less absorbent ones, caused the vaginal mucous membrane to dry up, and bits of the tampon sometimes remained in the vagina after it was removed—all of which facilitated a major staphylococcus infection. It was later found that toxic shock syndrome was caused by an exotoxin produced by *Staphylococcus aureus.* Certain kinds of highly absorbent materials (no longer used in tampons) take magnesium from the body, producing an environment favorable to the bacterial production of toxin.

However, there is a form of staphylococcus that is resistant to antibiotics, and its recent history is studded with the names of the major antibiotics to which it has become resistant. The development of this resistance can be summed up as follows. In 1952, 100 percent of staphylococcus infections could be cured by using penicillin; in 1982, only 10 percent of staphylococci were still sensitive to penicillin. Moreover, in 1980 a form of staphylococcus resistant to a major antibiotic, methicillin, was reported. It was given the now famous acronym MRSA (methicillin-resistant *Staphylococcus aureus*). Fortunately, there was the effective drug vancomycin—until, in the 1990s, forms of staphylococcus resistant to it (VRSA, vancomycin-resistant *Staphylococcus aureus*) appeared. Since then, press reports, especially in the United States, use the terms "superstrains" or "superbugs" to designate resistant germs, and the various pharmaceutical companies are more than ever engaged in a mad race to discover the antibiotic of the twenty-first century. When drug manufacturers announce a "new" antiobiotic, the new drug sometimes acts in more or less the same way as existing drugs. This explains why the number of forms of resistance to antibiotics does not really coincide with the number of so-called new antibiotics. Moreover, the creation of new antibiotics entails an increase in the cost of treatment. For instance, vancomycin is an expensive antibiotic; prescribing it puts a burden on the budgets of hospitals in rich countries, and it cannot be used at all in the health care systems of poor countries.

### *Bacteria and human ecology*

In such situations we encounter once again the conjunction of bacterial biology and human ecology. The emergence of bacteria resistant to drugs is partly the result of bacterial biology. In the case of pneumococcal infections in pediatric clinics in Durban and Johannesburg, we noted the mutating bacterium's tenacity; its exceptional resistance to antibiotics enabled it to reappear regularly despite repeated attempts to eradicate it. The germ's expansion was such that in 1978 it caused one out of every two cases of pneumonia in Durban. In reality, this indestructible microbe had already appeared ten years earlier in Papua New Guinea. Moreover, its residence in South Africa was a prelude to its spread to Europe and the United States. It was seen that a bacterium could acquire the genes for resistance from another bacterium, in the form of plasmids, the "Frisbees" mentioned in Chapter 2. This exchange of genetic material thus represents a form of bacterial sexuality: conjugation, an operation in which close contact between membranes permits the transfer of fragments of DNA. For Durban pneumococcus, it was not a matter of sexuality, the microorganism being little attracted by conjugation, but rather of a kind of gluttony that resulted in a transformation, the streptococcus in question revealing itself to be capable of ingurgitating long strands of alien DNA. In this way, it incorporated the know-how necessary to resist antibiotics, an ability conferred by the DNA it had received from a resistant bacterium.

In addition to these biological aspects connected with the private lives of bacteria, it has been demonstrated that bacterial resistance is more likely to emerge in socially disadvantaged environments. Poverty encourages self-medication, and in poor countries people often use out-of-date, inappropriate antibiotics, which they sometimes acquire from unprofessional sources or by fraud. All this leads to an insufficient, ill-suited use of antibiotics that not only fails to produce a cure, but also ends up selecting resistant forms of the bacteria. Similarly, in wealthy industrial countries, over-consumption and inappropriate use of the most recent antibiotics also leads to the selection of resistant bacteria. The association of poverty, lack of hygiene, and resistant bacteria explains the extent and the gravity of intestinal infections in poor countries and in ones that are less poor but have inadequate sanitary facilities. Where people lack toilets, where proper evacuation of human waste is not possible, or where the absence of running water facilitates the contamination of wells and reservoirs, people fall into the cycle that leads from the contamination of drinking water and food to diarrhea and then to further contamination, and so on. It is estimated that worldwide, since the 1990s, more than three million children under the age of five have died every year from diarrhea. Bacteria that are resistant to penicillin and that cause diarrhea have been known since the 1960s.[31] In her book *The Coming Plague,* Laurie Garrett tells the exemplary story of a Hopi Indian woman living on a reservation in Arizona. In 1983, the woman had a serious case of *Shigella* diarrhea that required hospitalization. The bacterium involved proved to be resistant to major antibiotics. How did this dangerous mutation appear on an Indian reservation? Simply because for the past three years, the Hopi woman had been using antibiotics inappropriately to treat a persistent urinary infection. By consuming antibiotics for months on end, she had made herself into a living incubator for the generation of a new, resistant strain of *Shigella*. Although the relevant hygienic actions were taken, by 1987 21 percent of the *Shigella* infections among Hopis and Navajos in Arizona were caused by the resistant strain that had appeared in 1983, and 7 percent of the cases of diarrhea in the United States as a whole were blamed on it.

Human consumption of antibiotics is not always intentional, because most of us absorb antibiotics without knowing it by eating meat and eggs. This story begins with the early stages of the production of antibiotics. The factories that isolated the molecules scientists were interested in (essentially penicillin) produced proteinic waste from the mold used. This material was considered a kind of vegetable extract, and was used as feed for pigs and chickens. The result seemed sensational: the animals grew and gained weight very rapidly.[32] Pigs and chickens reached adult size and weight in a short time—that was the benefit for farmers who raised them, but it transformed the animals into consumers of antibiotics present in the new feeds that were derived from the pharmaceutical industry. The generalized and unregulated consumption of antibiotics by animals results in antibiotics finding their way into foods consumed by humans, such as meat, eggs, and milk. All this encourages, of course, the emergence of resistant bacteria. Systematic therapeutic and preventive use of antibiotics in meat and dairy farming has made matters worse. Cases of bacteria resistant to antibiotics in animals intended for human consumption were reported in the United States, Holland, and Spain as early as the late 1980s and are now increasingly numerous. The presence of dangerous germs on farms can cause problems for humans even if they are very far away. In 1983, there was an epidemic caused by *Salmonella newport* in Minnesota, North and South Dakota, Nebraska, and Iowa. After lengthy investigation, it was discovered that the disease was connected to the consumption of hamburgers made from beef produced on a ranch in South Dakota. On this ranch, inappropriate use of antibiotics had encouraged the appearance of a resistant form of salmonella, which produced harmful effects when it was eaten in the form of hamburgers by people who were already taking antibiotics for medical reasons. In these cases, *Salmonella newport* found in the person eating the contaminated meat a site empty of bacteria where it could proliferate without competition.

In discussing the serious problem of the transmission of infections through foods, we mentioned the bacterium *Escherichia coli* 0157:H7 as a new, particularly dangerous arrival on the horizon of digestive diseases and infectious diarrheas. This bacterium—which is resistant to penicillin, the tetracyclines, and ampicillin—is present in the digestive tract of animals raised for human consumption. Contamination of *Homo sapiens* may take indirect and sometimes unpredictable paths.[33] Many public health officials are concerned about the consumption of antibiotics by animals raised to be eaten by humans. Despite alarmist statements and articles, things are not going to change any time soon. For example, the FDA authorized a concentration of antibiotics in cows' milk that is one hundred times higher than the previous limit. In many countries (though not in Sweden), combinations of antibiotics are used to promote the growth of livestock. Bacteria exchange genes, and the transfer of resistance (or its acquisition through the selection caused by antibiotics) is often accompanied by a transfer of virulence as well. In addition, genetically modified plants used in animal feeds include species that have been equipped in the laboratory with gene that resists antibiotics. It is easy to imagine the use that could be made of this by intestinal bacteria in animals eating this feed. Genes produced by genetic engineering can be found in the digestive tract and even in the blood of an animal after ingestion.[34] Without entering into the violent debate concerning the pertinence of using genetically modified plants, we should recall that this kind of modification has been blamed for spreading resistance to antibiotics.

### ANTIVIRALS, LIKE ANTIBIOTICS . . .

Drug makers produce far fewer antiviral molecules than antibiotics, for several reasons. First, it is much harder to create effective antiviral drugs; second, research in this area is more recent; and third, until the emergence of AIDS, the profits to be earned from antivirals may not have seemed great enough to generate much interest among pharmaceutical laboratories.

The first viral disease that attracted the attention of pharmaceutical firms was genital herpes, which in the 1980s was beginning to cool the ardor of the sexual liberation of the 1960s and 1970s.[35] For herpes, acyclovir was the equivalent of penicillin; it enabled physicians to cure the disease's main complication, herpetic encephalitis. In practice, it cures the symptoms but does not eradicate the virus, which has the unfortunate ability to remain sheltered in nerve cells and then reappear at the slightest weakness in the immune system or when the patient stops taking acyclovir. By the early 1990s, it was clear that there were herpes viruses that were resistant to acyclovir. The subsequent production of another antiviral drug, Ganciclovir, resulted in the emergence of multiresistant forms of the virus.

Patients sick with AIDS are often infected by viruses in addition to HIV, and they involuntarily participate, as in the case of tuberculosis and other bacterial illnesses, in the emergence of multiresistant forms of viruses. Among the latter are the herpesvirus and cytomegalovirus. There are now antiviral treatments for AIDS. Given our experience with earlier antivirals, it seems highly likely that forms of HIV resistant to the antivirals currently used will develop. HIV resistance has already been reported regarding azidothymidine or AZT, which is now an "old" anti-AIDS drug. Knowing that viruses in general and the AIDS virus in particular are capable of rapid mutation, we can easily imagine that the search for new drugs will be ongoing and that we are far from having found the definitive molecule that would provide a panacea for viral infections.

### RESISTANCE TO PARASITES

When DDT was first used to destroy the *Anopheles* mosquito as part of the campaign against malaria, people were all the more enthusiastic because the spraying of DDT also killed flies, fleas, lice, cockroaches, and other pests. Then DDT gradually began to lose its effectiveness against the mosquito, and even the flies reappeared, to the point that people expressed doubts about the quality of the insecticide used. In fact, the mosquito had become resistant, and calmly reestablished itself in the areas that had been treated with insecticide. This "refractory behavior" was observed by the experts on an ad hoc committee when they reported their results at every meeting. In the early 1970s, it was clear to officials at the WHO that there was no hope of eradicating malaria, given the resistance developed by the protozoon and the mosquito involved, and that another method for fighting the disease had to be found. Another complicating factor was that malaria was being "imported" with increasing frequency, in proportion to the growth of international travel. We speak of "imported" malaria when a person carrying the parasite settles in a previously uncontaminated area where mosquitoes are present to serve as vectors for the disease. In addition, malaria is seen with increasing frequency in urban areas, because of the water "reservoirs" mentioned in Chapter 2. When urbanization is combined with resistance, we see a situation comparable to that in India, where at least 130 towns have been attacked by "urban malaria."

## THE COST OF RESISTANCE

In developed countries, the major public health consequence of the emergence of resistant and multiresistant bacteria and viruses is a significant and unceasing rise in the cost of treatment. The chief victims of this increase are people living in poor countries, for whom the use of certain therapies is impossible. A single statistic can illustrate this situation: in 1990, Japan's per capita spending was, on average, $412 a year on medical expenses, whereas in Mozambique, the corresponding amount was $2. It might, therefore, be concluded that the simplest solution would be to distribute to poor countries the drugs most essential for their people's health, depending on the epidemiological situation in the country in question. However, one question immediately arises: who will pay for this? Neither the budget of the WHO, nor even that of the World Bank, would suffice. Even if these drugs were provided free of charge, that would not ensure that the people who need them most would have access to them. The example of ivermectin illustrates the situation faced in many countries and their inability to provide their people with minimal health care structures. Ivermectin, a drug developed by Merck, is used to treat onchocerciasis and to avoid the loss of vision caused by the disease. Onchocerciasis (also known as "river blindness") is a filarial disease caused by a parastic worm, *Onchocerca volvulus.* The disease produces symptoms (itching), and when the parasite penetrates the eye it can cause blindness. Onchocerciasis is one of the major causes of blindness worldwide. It is found chiefly in equatorial Africa and in Latin America. Considering the benefits of this drug and the poverty of most of the countries where onchocerciasis is endemic, Merck offered to give the WHO the drug for poor countries that needed it. It was up to the WHO to distribute the drug to some 120 million people who were exposed to the disease. The program began in 1981; in 1992, only 3 million of the 120 million who were at risk had received the drug, because of the extreme difficulty of distributing it. The effort failed not only because there was often a total lack of any health care structure, but also because of insurrections, coups d'etat, corruption, the absence of means of communication, and so forth.

### DESPITE THE MAELSTROM PRODUCED BY BIOLOGY AND HUMAN BEHAVIOR, CAN AN INFECTIOUS DISEASE DISAPPEAR SPONTANEOUSLY?

Whether they are emergent, resurgent, or resistant, the infectious diseases we have mentioned all show a general tendency to spread, or at least to persist. Whether their reservoir is human, animal, or vegetable, microorganisms have a remarkable ability to maintain themselves, usually concealed somewhere, apart from any epidemic, infectious disease, or other form of outbreak. Take, for example, the case of diphtheria. Between the two world wars, diphtheria was a contagious infectious disease that chiefly affected children under the age of ten and could be extremely serious. With the advent of vaccination and suitable modes of treatment, this disease "disappeared" in Europe and North America. Generations of medical students studied the symptoms of diphtheria only in textbooks. However, all it took was a breakdown of the social structure in the former USSR for us to see the disease make a comeback in an epidemic and fatal form. Once again, the bacterium had not been eradicated; far from it.

Is it also possible for a microorganism affecting *Homo sapiens* to disappear as a result of its own spontaneous, wholly natural evolution? The answer to this question is probably yes. First of all, many living species, including microorganisms, have vanished without any human intervention. It is generally agreed that since the creative explosion that occurred in the Cambrian period, more than 99 percent of species then extant have become extinct. However that may be, it is likely that since the appearance of *Homo sapiens* some bacteria have been eliminated. Take for example the disease called "sweating sickness," or "English sweat," which was extremely dangerous before it spontaneously disappeared. On August 22, 1485, the Battle of Bosworth Field put an end to the War of the Roses and made Henry Tudor king of England. Perhaps in relation to this conflict, a month later an epidemic of sweating sickness broke out, quickly reaching its apex in London before spreading to the rest of England, but sparing Scotland and Ireland. The sweating sickness struck young, vigorous men, causing respiratory distress preceded by severe headaches and peculiarly malodorous sweating. Victims became delirious and fell into a state of semiconsciousness; many of them died. This disease terrorized England and then vanished for unexplained reasons; it existed only between 1485 and 1552. In fact, there had been no sign of it between 1486 and 1507, when it reappeared in London. It appeared again in 1518, when it reached Calais but affected only the English there. We know virtually nothing about this disease; was it a form of dengue or other viral disease? In many respects, the sweating sickness was a very strange malady, particularly in its limitation to the English. It is perhaps the only epidemic disease the historical existence of which is not in doubt and yet which spontaneously disappeared. Was its extinction final, or will we see a resurgence of English sweat? The future will tell, but it would be presumptuous to speculate about an eternal extinction of this strange sickness. The wisest attitude is to assume that in principle an infectious disease is never eradicated. The long absence of English sweat, which is sometimes said to be a permanent disappearance, is not necessarily a final extinction.

## RESISTANCE, AN ANTHROPIC PHENOMENON BETWEEN NATURE AND CULTURE

At the beginning of this book, we said that an epidemic cannot be explained by a microbe alone. We have seen that abiotic, ecological, and human factors contribute to outbreaks of epidemic diseases. Nothing prevents us from thinking that such elements could provoke a resurgence of English sweat or smallpox. Resistance to antibiotics, antivirals, and drugs used to treat malaria can be illustrated by the third loop in the emergence of diseases, the loop of resistance and reappearance (Figure 3-1).

The loop of resistance reflects an essentially human phenomenon. The role played by so-called natural phenomena is secondary and sometimes nil. *Homo sapiens* is also part of nature; we belong to nature in our somatic and mental wholeness. As Moscovici puts it, art and technology do not constitute an antinature. An anthropic phenomenon is not necessarily antinatural. The resistance, emergence, and re-emergence of microorganisms, even when they use human phenomena as a support, are nonetheless natural events. Denunciations of them as "not natural" belong to the domain of morals, not to that of biology: human intelligence and knowledge are the outcome of natural phenomena. Our current mental faculties lead us to increase our ability to produce, construct, invade, and modify our environment. Since Neolithic times, the natural tendency has been to

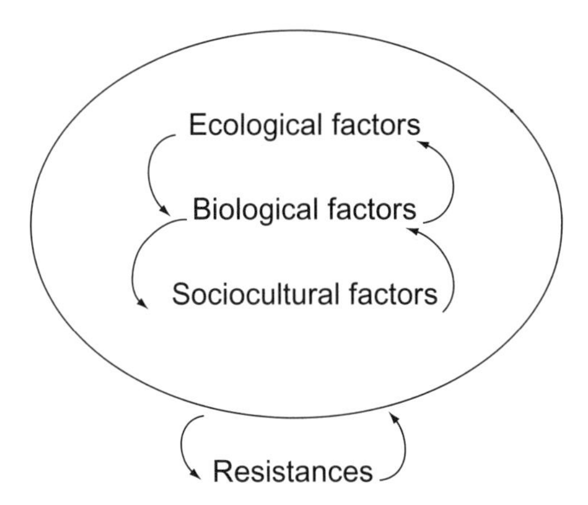


**FIGURE 3-1** The third loop, the loop of resistance.

increase the amount of cultivated, productive land in order to feed a more numerous family, a larger group, a more extensive country. The tendency was (and remains) toward the growth of an agricultural economy. It is worth noting that the most frequent geographical origins of infectious diseases—whether old or new, emerging or re-emerging—lead us back to the origins of humanity and to the beginnings of human technology, that is, to agriculture. Many diseases caused by microorganisms that are or have been problems on a global scale first emerged in the tropics, often in Africa. Africa is, of course, considered the cradle of humanity, and it remains the mother lode and the part of the world where many major pathogens are prevalent. Against some of these diseases of tropical Africa, *Homo sapiens* has not yet developed complete immunity, only partial protection (as in malaria) or none at all (as in AIDS). Similarly, the protection provided by the human immune system against most tropical diseases caused by parasites is inadequate. In addition, an intermediary host is usually available to serve as a vector and thus offers the parasite shelter. Finally, the parasite's biology is connected with the water that is usually required for the vector's life cycle. It is interesting how the African context has always favored the infection of human beings by microorganisms from which they can never completely free themselves. Might the constant infections have perhaps encouraged the migration of *Homo sapiens* to other, less hostile areas, including the future Europe? In its modern form, agriculture, the first human technoculture, has encouraged the emergence, resistance, and re-emergence of various diseases—essentially by modifying the environment, but also by weakening the host.[36] Malaria, schistosomiasis, filariasis and diseases caused by arboviruses such as Japanese encephalitis are all linked to agricultural expansion. Tropical diseases show how socioeconomic, demographic, and religious behaviors are involved in the emergence of diseases caused by microorganisms, that is, of the elements of the domains of biocultural anthropology and ethnomedicine. Brinckmann has provided a good summary of the way the various actors in the process amplify the mortality and morbidity of diseases associated with the expansion of agriculture. Two elements are regularly found at the origin of emergences and/or resistances: deforestation and irrigation. Both are intended to increase the area cultivated and the yield per acre. The geographical expansion of the cultivated surfaces facilitates contact with animal diseases present on the territory gained, and this results in *Homo sapiens* being exposed to diseases that can be contracted from animals. The conquest of new territories that may be inhabited by "vermin" and the desire to increase yields encourages the use of pesticides, herbicides, and other synthetic products that aggravate an unbalanced biocenosis already compromised by the simple presence of humans. In practice, it is common to say that here we see a factor of ecological imbalance, but is it in reality an imbalance or the creation of a new balance? The new situation may facilitate the emergence, and/or the resistance and/or the multiplication of vectors, intermediate hosts, and so forth. If, by hypothesis, we suppose that the movement toward a new agricultural economy ends up producing a tangible result, then labor will be required, and immigrants will arrive—and they will normally be less well prepared than the natives to deal with the local infectious agents. Immigrants are often disadvantaged people who live in marginal and unsanitary conditions and thus meet all the conditions for encouraging the prevalence of infectious diseases. The model presented here corresponds to reality in many developing countries. It has one aspect that at first seems paradoxical: according to a dogma that is virtually equivalent to a paradigm, "development" is synonymous with social progress and improvements in public health that are accompanied by a decline in infectious diseases. This commonplace dogma must be qualified, since Lyme disease, for example, followed the economic development of New England associated with the agricultural and industrial growth of the former English colony. The emergence of *Escherichia coli* 0157 : H7, which has by now infected millions of people, is clearly linked to technico-industrial "progress," particularly in the area of food production, and in general to everything connected with modernity. Thus, in a developed country, emergences can be all the more numerous because people are trying to promote new economic activities, especially

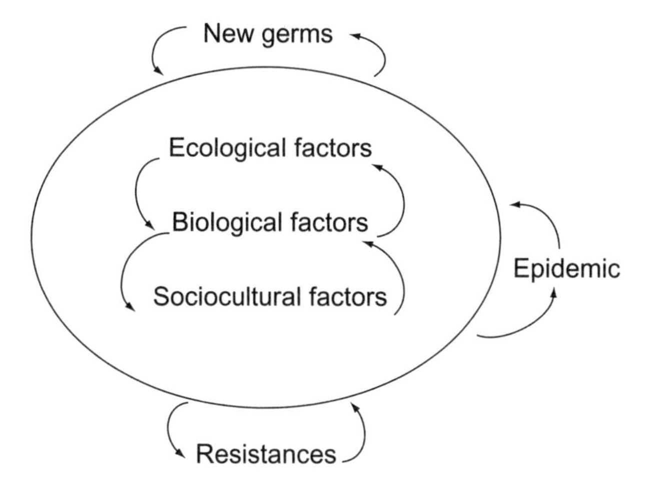


**FIGURE 3-2** The overall loop of epidemics.

agricultural activities. In 1996, infectious diseases represented 33 percent of the causes of mortality in the world; but there was a grotesque difference between the developed world, where 1.2 percent of the mortality resulted from infectious diseases and diseases caused by parasites, and the developing world, where 45 percent of the mortality was due to these same causes. In countries experiencing severe problems, the WHO fights fiercely, and not without difficulty, to limit the pernicious effects of these various diseases, but wars, insurrections, and coups d'etat are constantly hobbling the efforts of caregivers in the field. Despite various difficulties, the WHO reported the following results in 1997: (1) between 1995 and 1996, a decline in the rate of the prevalence of leprosy, from 2.3 to 1.7 per 10,000 residents; (2) the continuation of the program of protection against onchocerciasis that began in 1974 and that by 1996 had protected 36 million people;[37] and (3) in India, the vaccination of 120 million children against polio on a single day in 1996.

The tropical model as described by Brinckmann merely reproduces a general model in which, with the exception of a few details, all the factors described in the preceding loops are present; epidemics, resistances, and new germs are brought together in such a way that they are all included in a single model or loop. The conditions for resistance are not much different, except for small details, from those of emergences and re-emergences. The whole affair could thus be represented by an "overall loop of epidemics" (Figure 3-2).

Taking into account what we have observed in relation to the histories of epidemics, the emergence of new diseases, the return of old diseases and the persistence of others, we can emphasize certain common elements. These elements are noteworthy insofar as the appearance of infectious mechanisms connected with human behaviors is concerned, whether this is a matter of behaviors at microlevels or at macrosociological levels. Similarly, there are analogies between the reactions of human groups affected by infectious diseases. These reactions can be integrated into universal models, to which must be added attitudes that have cultural specificities related to each society. Of course, it is not a question of omitting ecological and biological factors, whether the biology in question is that of humans or of microorganisms. In other words, the knowledge mentioned in the preceding pages encourages us to move beyond the concept of the host-microorganism relationship that might at first seem obvious and inevitable. First of all, we should note that the biological relationship between host and microorganism has to be taken into account as a whole, at a different level of integration. It is identical with the machinery of sociological and cultural behaviors. We must emphasize that the emergence of an epidemic of a resistant germ or the re-emergence of a pathogen that had been considered practically eradicated do not result from biological causes alone. Biosocial and biocultural approaches demonstrate without the shadow of a doubt the roles played by social and cultural factors in the birth of an epidemic, just as epidemics favor the appearance of sociocultural factors. We will not repeat here what has been said about animals, microsociological events such as sexual practices, and so forth, that contribute to the processes of emergence. We have already drawn attention to the role of macrosociological behaviors such as deforestation, ecological damage, migrations, and urbanization in the mechanisms generating attacks by microorganisms on human groups. Sociological reactions are often common to various societies and different epidemics. Among these behaviors we note the usual flight away from the site of the epidemic and the practice of therapeutic measures, whether rational (asepsis, quarantine) or irrational (appeals to the supernatural or to religion). However, these reactions also often lead to conflict between or within groups. Fear can lead to aggression; the search for guilty parties, for scapegoats, leads to ostracism and rejection.[38] Spiritual or religious repercussions are usually also seen when epidemics or infectious cataclysms wane. In fact, none of the psychological, ecological, or biological elements can or should be studied in isolation. An observer of major infectious phenomena worldwide gains the impression that *Homo sapiens* always lives in a state of provisional freedom in which pathogens constitute a constant threat. Consider, for example, the emergences mentioned on the WHO's Web site on any day chosen at random.[39]

The relationship between *Homo sapiens* and infectious agents is in a permanent state of armed peace, a situation of complex equilibrium that can be disturbed by the smallest change. Biological and cultural factors that make it possible to safeguard an often precarious equilibrium constitute the immunity of populations.

CHAPTER 4

# THE IMMUNITY OF *HOMO SAPIENS*: *From the Individual to the Group, from Biology to Culture*

Like our distant ancestors, we are surrounded by microorganisms that may or may not be pathogenic but are natural, inevitable inhabitants of our ecosystem. We carry microbes with us, and we serve as their hosts. Each of us is a veritable zoo of various microorganisms that constitute a potentially hostile environment against which we must protect ourselves. The available means of doing so are found in the domains of biology and culture. Here, the term "culture" refers to the concepts, means, and behaviors human groups use to control their ecosystem. If an individual's defenses fail, that is tragic for him alone, but if a whole human group's defenses fail, that is a catastrophe for the population involved.

## THE IMMUNITY OF POPULATIONS

What matters for a population is not whether a given individual falls victim to an infection but whether the group survives despite the casualties, for it is the group that constitutes the biological unit. It is as if a kind of de facto solidarity exists among individuals, an immunological altruism, especially during epidemics; some individuals die, while others survive, thus ensuring the group's continued existence. It is no longer a question of individual immunity but rather of the federation of each individual's immunities in a kind of overall immunity of the population, in an immunological supersystem that may be called the immunity of populations. The latter is an emergent phenomenon, a collective immunological heritage that is subject to the vicissitudes of environment and culture. Humanity has always suffered, and will continue to suffer, from pathogenic agents. The immunology of populations, that is, the study of populations' immunity, should allow us not only to shed light on epidemics from a different point of view but also to do a better job of defining problems of public health connected with microorganisms and their harmful effects.

Ecologists seem to neglect the role of microorganisms as components of the natural environment; whether or not they are addressed to academics, works on ecology seldom mention events (such as epidemics) or factors (microorganisms) that are potentially damaging for humanity. The human immune system is an essential factor in our relations with the external environment because it is confronted by microbial aggressors. The immunology of populations therefore does not concern solely the relationships between humans and microorganisms; it is not synonymous with the history of epidemics, but instead concerns the species' immune system as the agent of various kinds of interactions with the environment, including those with nonmicrobial elements of our ecological niche.

A discussion of the immunology of populations presupposes a general knowledge of immune processes, that is, of the means organisms use to protect themselves against attacks by microbes. We use the term "microbes"[1] here to refer to microorganisms, including bacterial, viral, and parasitical agents of all kinds. The immune system, which is simultaneously cognitive and defensive, has the ability to remember. Descriptions of immunological phenomena elicited by microbial intrusion and of the immunological basis for these phenomena often make use of military metaphors referring to the "defenses," "killer cells," "rejection reactions," "surveillance," and "detection of intruders" that oppose "targets," "aggressors," "invasions," and so forth. However, discussion of the immunity-microorganism relationship from a strictly conflictual point of view is questionable in many respects.

In its biological unit, the human group, the immunity of human populations represents the emergence of a general response that itself arises from the immune response of each individual. The development of a microbial attack is related to the destructive potential mobilized by the totality of individual responses expressing the immunological status of the population with respect to a given pathogenic agent. The more individuals there are within a population who respond to the attack, the greater the population's immunity. In the course of a prolonged struggle (years, decades, centuries), both the defenses of the population under attack and the microbes themselves may coevolve. The immunity of populations is a dynamic state that varies with time. It may undergo adaptive phenomena, possibly on a neo-Darwininian model, but probably also randomly.

Let us examine first the immune system on the individual level. The term "system" is used here in the sense defined by Morin: "An interrelation of elements constituting an entity or unit as a whole." The military images of immunology illustrate the ability of the individual's immune system to recognize foreign substances, marked as not-self or antigenic, on one hand, and his own tissues or self on the other hand. That is the "classic" definition of the immune system as an extremely complex biological apparatus capable of distinguishing between the self and the not-self. A more modern conception defines the immune system's machinery as a totality of agents that interact not only with each other but also with other elements in the internal or external environment. Many theorists do not grant the "biological self" a fixed identity, because, for example, the human genome contains DNA of viral origin. Everyone's "self" includes exogenous elements. Understanding immune response mechanisms is not easy for someone not familiar with immunology, but the Manichean conception of the self opposed to the not-self makes things easier. It is possible to distinguish defensive processes with rudimentary systems of recognition from those whose instruments of detection are more elaborate. To this distinction between the cognitive procedures of the immune system corresponds a difference in the kinetics of response: immediate, with the help of elementary means, and delayed, when it relies on complex mechanisms. An account of how our organism protects itself from microbes thus takes us back to the notions of the self and the not-self, that is, to a very specific form of biological individuality. The terms *not-self* and *self* belong more to the domain of metaphor than to that of scientific definition. For the immunologist, the opposition between self and not-self is habitual and convenient when formulating concepts. The not-self represents the "foreign," the alien element against which the self will produce a response by mobilizing defense mechanisms. The markers of human polymorphism, such as blood groups, make it possible to take an organic, tissue-based, molecular approach to the self and show that the probability of finding two subjects with identical biological traits (taking into account only the most common biological characteristics) is one in a billion. An individual's biological identity can attain a virtually absolute uniqueness; for example, in the case of one person studied by Jean Salmon, director of the French blood bank in the 1960s, the characteristics were so peculiar that the probability of finding an identical subject was one in a hundred million billion, that is, a number far larger than that of all the humans who have existed since the beginning of humanity. Thus with the help of certain markers, the biological self can be defined in a way that is precise but always contingent on the markers selected.

### *The skin and immediate immunity*

"Of all the sense organs, the skin is the most vital . . . unless most of the skin remains intact, one cannot survive".

In advance of elementary defense processes, mechanical barriers provide a very important kind of protection. The skin and mucous membranes constitute an obstacle that microbes usually cannot overcome. These barriers combine integuments and their allies, such as tears, mucus, and bronchial cilia, in order to form a mechanism for evacuating impurities. In the event of a microbial attack in which the skin has been penetrated, it is the instruments of natural immunity, those of the immediate defenses, that are called into action. The latter have a limited ability to discern intruders: their recognition of the not-self is rudimentary. Among the immediately available elements, the complementary system constitutes a redoubtable weapon against microbes. As a whole, it consists of a family of thirty factors soluble in blood or associated with cells. With prodigious efficiency, it attracts cells that devour microbes, the so-called macrophages, which engulf the microorganisms and then destroy them by producing toxins. The phagocytes, represented by polymorphonuclear cells and macrophages, discern on the surface of the microbes the molecules characteristic of microorganisms. Phagocytes are present throughout the organism; as vigilant guardians, they circulate in the blood, but they are also found in tissues, the principal viscera, muscles, and so forth. When microbes are consumed by phagocytes, the lymph organs (the spleen and lymph nodes) become battlefields between the self and the not-self. This explains the plague victims' swollen lymph nodes (in which the cells of the plague bacillus have established themselves) and the enlarged spleens of individuals infected with the malaria parasite. If the invading microorganism is a virus, it will find itself opposed by very special cells that are also "already prepared" to act in advance of any infection, natural killer (NK) cells.

We have just seen how immunity's foot soldiers, prompt to respond but lacking in discernment, act within minutes to prevent an invasion by pathogenic agents. This vanguard does not act alone but is accompanied by soluble mediators, the proteins produced by phagocytes, in particular macrophages. In other words, the preceding cellular agents are reinforced by substances produced through their own secretions. These mediators are known as cytokines or interleukins; the main ones are interleukin 1, interleukin 6, interferon, and interleukin 8, as well as the tumor necrosis factor (TNF). The interleukins of the immune system provoke inflammatory reactions that are usually beneficial, and they facilitate defense processes by increasing blood flow to the site (in the case of an abscess, for example) and the movement of phagocytes toward the infected tissues. We have already seen (Chapter 3) that abnormally high rates of production of these soluble mediators are responsible for toxic shock syndrome.

Natural immunity has no memory, whereas acquired immunity does. For example, someone who has been vaccinated against tetanus is immunized specifically against that disease; he was not immune before the vaccination, but he is afterward. His immunity is acquired. The immunization produced by vaccination protects him against tetanus and only against tetanus; acquired immunity is specific. Any new contact with the vaccine will produce a reaction mediated by the immunological memory and characterized by speed and intensity. Acquired immunity is established after stimulation by the not-self and produces a specific response while at the same time retaining an imprint, a memory of the first contact with the not-self concerned. Acquired immunity, which is extraordinarily complex, involves a unique kind of cells called lymphocytes. Specific immune response, including its memory and its regulation, exists only thanks to lymphocytes. Acquired immune response is called clonal, that is, the lymphocytic reaction to each kind of not-self is supported by a group of cells that are identical with each other, specific to this not-self and to no other. Lymphocytes have the property of being able to recognize not only the not-self but also the self, and this is a characteristic of acquired immunity. Lymphocytes, which are so valuable for defense against microorganisms, include two different populations, T lymphocytes and B lymphocytes. The former derive their name from the fact that they have been "trained" by a major organ of the immune system, the thymus, whence the label thymodependent lymphocytes or T lymphocytes. B lymphocytes arise from the bone marrow. T and B lymphocytes, usually acting together, make an efficient coalition against microbes.

The major histocompatibility complex, or MHC, is known as a genetic or molecular organization, a fundamental element in graft rejection (an artificial situation resulting from medical intervention), but it is in fact far more than that. The complex constituted by the MHC and the antigen is detected by T lymphocytes. A given T lymphocyte picks out, "sees," only one antigen, by means of antigen receptors. The totality of T lymphocytes includes a large variety of receptors, as many as hundreds of millions, thus constituting a formidable capacity for recognizing different antigens. This faculty represents the repertory of T lymphocytes. Similarly, each B lymphocyte "sees" a single antigen to which it is specific. The membranous receptor of each B lymphocyte consists of a molecule of an antibody acting as a receptor for antigens. In the presence of antigen the B lymphocyte becomes an antibody factory. The antibodies it produces are extremely diverse, to the point that they have been subdivided into various kinds of immunoglobulins.[2] Antibodies may be present in the bloodstream, organ tissue, and the mucous membranes, where they provide an immunological coating. These kinds of antibodies correspond more or less to the place where the immunoglobulin is active: G and M immunoglobulins in the blood serum, A immunoglobulin in the mucous membranes.

## THE PRINCIPAL BIOLOGICAL MARKERS OF THE SELF IN THE IMMUNOLOGY OF POPULATIONS: THE MHC AND THE ABO SYSTEM

We owe the discovery of the first MHC type[3] to Jean Dausset, who was to receive in 1980 the Nobel Prize for Medicine for his discovery, and who has recently published an account of this work. It is as the operator of immune response that the MHC interests those concerned with the immunology of populations.

### *The HLA (human leukocyte antigens) system*

At the beginning, the HLA molecules present on human cells were considered as products responsible for graft rejection; it is the reason why they were named "antigens." Now we know that they are more than that. Besides the fact that they are involved in the process of immune response, they are a sort of fingerprint of our cells. Each individual possesses its own HLA formula.

With the exception of red corpuscles, the HLA system's molecules are present on the surface of all human cells. The characteristics of an individual's HLA are genetically determined; they are inherited from his parents. In every human, the HLA molecules are diverse and, as a result, the identity of any given individual is practically unique. To be sure, we may share one, two, or—rarely—three HLA alleles with someone else, but almost never eighteen or twenty. The extent of the diversity varies; it is relatively great in Europe, for instance, but quite limited among the

Indian populations of the Americas (Amerindians). Moreover, certain characteristics occur more often than others. Since the HLA's products are associated with important biological phenomena, it is not surprising that its diversity is involved.[4]

### *The ABO system*

In 1900, Karl Landsteiner, who was both a physician and a chemist, showed that the human race is divided into four major groups called A, B, AB, and O in the ABO system. Landsteiner's discovery inaugurated the era of blood transfusion. No serious work in this area had been done earlier; it was all made possible by the discovery of the ABO system. This system was described as an erythrocytic system (i.e., a system of red blood cells), but its products are present on all tissues. The transmission of ABO characteristics can be described in terms of Mendelian genetics by taking the O gene as recessive and the A and B alleles as codominants. In view of these data, identical phenotypes (groups) correspond to different genotypes. ABO characteristics are associated with certain illnesses, including infectious pathologies. Nonetheless, the biological function or functions of these molecules (if they have any) remain obscure.

## BIOLOGICAL MARKERS AND INFECTIOUS DISEASES

"When I began practicing medicine, measles was one of our preoccupations. Between 1900 and 1910, measles killed about one million people. During the same period, in France it killed more than thirty thousand children in eight years . . . thus to fight measles was to fight the contagious and epidemic illness most dangerous for French children. . . . But we were well aware that the true way to fight measles was vaccination."

The preceding lines were written by the great pediatrician Robert Debré, who lived through a magnificent period of transition for physicians and patients during which antibiotics were discovered. He saw the disastrous effects of many infectious diseases and was aware of the potential danger they represented—and continue to represent, as was pointed out in a 1992 in *Nature* in which the author noted that "infectious diseases are not banished."

For the student of the immunology of populations, it is usually difficult to identify a marker of vulnerability or resistance to most of these diseases. For example, it is hard to tell whether the evolution of syphilis was more deleterious among certain human groups than among others, nor is there any trace of a solid biological and genetic marker for tuberculosis; as for leprosy, the question is debatable. The same goes for yellow fever, although it is clear that in countries where the disease is endemic the natives show more resistance to it than do immigrants coming from countries where the disease is unknown. What should we say about the "English sweat" that appeared and disappeared without our really knowing how or why? Thus it is sometimes very difficult to demonstrate, in the human species, a connection between an infectious disease and one or more genetic markers. It may be easier to do so for cattle, for Mareck's disease in chickens, or among mice.

Nonetheless, there are cases in which investigations produce useful information, such as those dealing with schistosomiasis, an illness caused by *Schistosoma mansoni,* a parasitic worm also known as a blood fluke. In a study of forty-five families in northeast Brazil, Dessein clearly demonstrated the existence of a gene that controls the intensity of the infection. It has also been observed that, among Egyptian schoolchildren, serious forms of the disease were associated with the alleles HLA-A1 and HLA-B5. The relative risk of serious illness is high for children with HLA-A1, 18.5 for those with HLA-B5, and even higher for young patients with both A1 and B5. Among Japanese who present a related schistosomiasis, that associated with *Schistosoma japonicum,* the association with HLA is based on different alleles and reveals a complex mechanism in which one allele (HLA-DR2) is thought to be the indicator of a "positive" immune response to the parasite, whereas another (DQw1) is thought to help the illness establish itself.

## BIOLOGICAL MARKERS AND VACCINATIONS

Like the HLA system, the markers of natural immunity have been associated with how well an individual responds to vaccinations. Thus it has been noted that the association between two alleles (HLA-B8 and DR3) is accompanied by a weak response to hepatitis B vaccine. More precisely, in this case it is homozygotes who are weak responders, proving that a dominant gene suffices to produce a high-quality immune response. Vaccination against hepatitis B of normal subjects and subjects with renal insufficiencies has shown that the allele HLA-DR3 is more often found among non-responders to the vaccination. Here, the HLA alleles involved are the markers of a poor response. All information regarding individual vulnerability to hepatitis B is of the greatest interest, because worldwide three hundred million people are infected by the virus that causes this disease, and its presence multiplies by one hundred the risk of cancers originating in the liver. The connection between hepatitis B and the HLA system has not been verified by all researchers; one study of the vaccination of two hundred patients receiving chronic hemodialysis has shown that the group of weak responders to the vaccine included fewer subjects with HLA-DR2 than the control group, without an increase in the number of patients with HLA-DR3 among the nonresponders.

In other contexts, the search for a connection between response to vaccination and a biological marker has shown that the response—for example the response to streptococcus antigens—was not dependent on the HLA alleles themselves but rather on genes close to those of the human major histocompatibility complex (MHC). Similar investigations have been conducted during vaccination against measles, a disease that is sometimes very serious and is still common in developing countries. First of all, it has been observed that measles is particularly severe among certain ethnic groups, and that this severity is associated with HLA alleles, as in the case of the Zulus living around Durban, where the HLA-A32 allele, normally present in 2 percent of children in the control group, is found in 10.1 percent of those with severe cases of measles. The connection between HLA immune response and measles has also been demonstrated among Dutch children in the course of an investigation of 143 pairs of identical and fraternal twins. To this we can add studies showing that patients suffering from multiple sclerosis and having HLA-A3 and HLA-B7 alleles produce large numbers of antibodies against measles. Studies on vaccination against tetanus have provided less conclusive information; among Japanese students an association between a poor response and HLA-B5 has been observed, whereas another study of European families did not demonstrate a connection between response to vaccination against tetanus and the HLA system.

Dutch physicians have observed that among Dutch naval cadets who have never been vaccinated against smallpox, the response to the vaccine virus was connected with HLA alleles, and, more precisely, young men with HLA-Cw3 showed a weak immune response following vaccination.

Another vaccination that has been studied is that against *Haemophilus influenzae,* a bacterium that infects the upper airways in children and is the chief cause of bacterial meningitis in children between the ages of nine months and four years. While the vaccine produces excellent results among young Europeans, it has been observed that this is not always true for other populations. This failure of vaccination is not surprising among Amerindians, in whom infection with *Haemophilus influenzae* is ten to fifty times more frequent than among Caucasians; thus vaccination of Eskimo and American Indian children provided no protection. A more detailed investigation carried out in Arizona among Apache children between the ages of eighteen and twenty-four months showed that they produced fewer antibodies against *Haemophilus influenzae* than white Americans. The same Amerindian children were capable of a completely normal response to vaccination against diphtheria and tetanus. This aspect of defenses against influenza does not seem to be connected with consanguinity (a common recent ancestor) as has been shown by a study carried out among the Amish, an American religious group that lives to a large extent apart from the surrounding society and thus forms a kind of isolate.

To return to the Alaskan Eskimos, who have the world's highest rate of infection with *Haemophilus influenzae* type B, it is remarkable that many in these populations are born with a metabolic irregularity resulting from a deficiency in the enzyme 21-hydroxylase, an anomaly whose gene is part of the HLA complex. This condition provokes an illness known as congenital adrenal hyperplasia, which disturbs the metabolism of sodium but protects Eskimos from the meningitis associated with *Haemophilus influenzae*.

## THE IMMUNOLOGY OF POPULATIONS: ITS EVOLUTION MARKED BY ITS ECOLOGICAL NICHE AND BY CULTURE

### *The Conquista*

In 1518, the Spanish left Cuba and headed west; Cortés, partly through luck, arrived on the American continent in the year known as *Ce Acatl* in the Aztec calendar—the year in which the plumed serpent Quetzalcoatl was supposed to return. Cortés took almost a year to march from his landing point in Yucatán to the city of Tenochtitlán or Mexico City, which he reached on November 8, 1519. The Mexican capital fell on August 13, 1521; by 1522 half its inhabitants were dead, as a result of the war and infectious diseases. Epidemics were to strike Mexico throughout the sixteenth century, in 1531, 1541, 1564, 1576, and 1595. The *Conquista* is said to have led, directly or indirectly, to the death of fifty-six million people. The arrival of the Europeans is considered to have been a disaster for the health of the Amerindians, because researchers have found no trace of enormous deleterious effects produced by acute or chronic diseases in the pre-Columbian period. The Spanish and Portuguese conquest introduced horses, pigs, sheep, wheat, and cotton into the New World, and took back corn, potatoes, tobacco, and cocoa; but it also contributed to an exchange of microbes. Syphilis is thought to have been brought back from America, and European viruses followed the *conquistadors,* so that microorganisms are associated with what was perhaps the greatest ecological transformation in the history of humanity: the discovery of America.

The vulnerability of the Amerindians when exposed to European viruses is now thought to have resulted from the fact that these human groups were essentially composed of individuals who were weak responders to microbes brought in from the Old World. This vulnerability is classically correlated with the weak genetic polymorphism of these populations, as is shown, for example, by the fact that the great majority of them belong to the O blood group, even though more than 50 percent of the members of the Blackfeet tribe have type A blood. According to Ruffié, this monomorphism results from selective pressures and is not a matter of chance, which causes genetic deviation. In other words, the genetic peculiarities of the Amerindians result from the evolution of these populations under the selective pressure of their environments.

### *The original migration*

According to one theory, the Amerindians' genetic monomorphism is thought to have developed before the arrival of the Spaniards in the Americas. The true discoverers of the New World came from Asia, traveling on foot over the Bering Straits, which were then frozen solid, and leaving Siberia in at least four migrations across the straits. The first of these migrations took place seventeen thousand to twenty-five thousand years ago, the second between ten thousand and thirteen thousand years ago, and the third and fourth ten thousand and eight thousand and years ago, respectively. The origin of Amerindian populations is attested by biological evidence—the study of some of their genetic markers, such as the Gm alleles in the G immoglobulins shared by Asians and Amerindians. This biological approach is supported by studies that confirm the Asiatic origin of these various groups' languages. No one questions that Amerindians are genetically homogeneous, a fact that is clearly demonstrated for the ABO and HLA systems; the latter is characterized by an unusually weak polymorphism[5] that is classically correlated with the disastrous effects of epidemics.

### *The Mexicans' demise after the Spanish conquest*

Why did the Amerindians of Mexico die? Their genetic monomorphism was the cause and/or the result of the high mortality rate. Taking into account what we know about HLA molecules, it is not surprising

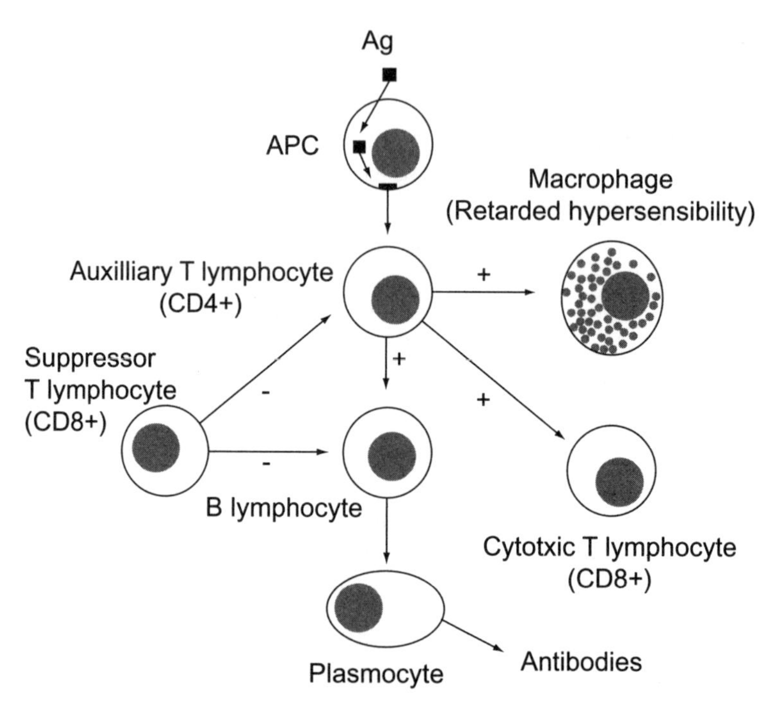


**FIGURE 4-1** The main players in the immune system. The antigen (Ag) is perceived by an antigen presenting cell (APC), which may be a dendritic cell or a macrophage. The antigen presenting cell "shows" the antigen associated with an HLA molecule to the T lymphocyte type CD4+. The latter will be the cell responsible for the specific immune response. This response may employ various means, such as the generation of killer cells (cytotoxic CD8+ lymphocyte T cells), an inflammatory reaction called delayed hypersensitivity, and the production of antibodies (Ac) by B lymphocytes transformed into plasmocytes. The immune response is regulated by suppressing lymphocytes that are also CD8+.

that the monomorphism of the MHC within a population is accompanied by limits on that population's ability to respond to microbes. The HLA molecules play a major role as components of tissue-related individuality and as molecular structures presenting antigens. In each individual, this diversity makes possible the presentation of all the antigens individually "shown" to the T lymphocytes that have developed discriminatory capacities (see Figure 4-1). In order for this to function, an HLA structure must be able to present several antigens. We see that not everyone has the aptitude, because of the nature of his MHC, to present all the antigens that his immune system might encounter in nature. In other words, our

“HLA keys” do not allow us to open all the doors of immune response. The inability to present these keys, and thus the immune system’s ability to respond to one or more antigens, represents gaps in the subject’s immune repertory. If these gaps concern microbial antigens, the subject will be vulnerable to the infectious agent concerned; as a nonresponder, he may fall victim to a serious and possibly fatal disease. Someone with this immunological handicap is penalized, but his human group will not be, if it contains individuals whose immune system includes molecules of the MHC that make possible the presentation of antigens. Consequently, within a population, the wealth of HLA diversity may result in a greater probability of responses to a broad variety of microorganisms, since the HLA system represents the group’s immune repertory.

To return now to the Amerindians, the usual view attributes their misfortunes to an inability to offer a strong response to microbial attacks that results from an absence of immune response or a weak ability to mount one. The cataclysmic epidemics that followed the conquest of the American continents are thus thought either to have been fostered by HLA monomorphism preceding the conquest or to have led to this monomorphism by eliminating the nonresponders, leaving as survivors only those whose polymorphism, which resulted from the powerful selective pressure, was diminished.

In fact, matters are still more complicated, since on the hypothesis that the disastrous effects of the epidemics were facilitated by monomorphism, the latter may have produced a poor response not because of the immune system’s cognitive deficiency, but rather simply because monomorphism allowed the virus to progress more rapidly within the population. According to this theory, the Amerindians’ vulnerability to infectious agents would be due not to an absence of immune response genes (HLA) but rather to the group’s genetic homogeneity itself. It is well known that in viral infections the disease becomes more serious when the virus is exchanged between members of the same family. In Africa, where vaccination against measles has not been widespread,[6] a child who falls victim to this disease is twice as likely to die from it if the virus was transmitted to him by a relative. It is as though the virus adapted to the MHC, to an individual’s HLA molecules, an adaptation that makes it more virulent among those who have the same HLA genes, whether or not they belong to the same kinship group. When the population is homogenous in terms of HLA genes, a virus that has adapted to a given set of genes will more easily find individuals whose alleles it “knows.” Amerindians with weak HLA polymorphism would thus have been victims of this homogeneity, the latter having existed before the Spanish invasion, when the ecological niche, lacking European microbes, allowed American populations to live without suffering major epidemics. In interpreting the preceding views regarding measles in kinship groups, the specific ways in which the disease is transmitted must also be taken into account.[7]

### *The epidemiological transition*

Studies of various vaccination programs among Amerindians have proven that these groups do not necessarily include a majority of weak responders. These communities' primary vulnerability to infectious agents must have existed before their conquest and they remain vulnerable because they are weak responders to certain vaccines. Nonetheless, this vulnerability should not be assimilated to a kind of paradigm providing an excuse. The study of American populations in subarctic regions shows that, thanks to socioeconomic progress, these communities, which between 1714 and 1821 experienced multiple epidemics of smallpox, measles, whooping cough, influenza, and tuberculosis, no longer have such a prevalence of these illnesses.

The poet and travel writer Alain Gheerbrant encountered the Yanomami tribe, then called the Guaharibos, in 1950. This group, which inhabited the Amazon rain forests, lived comfortably—according to their standards, of course—in an environment that most outsiders would regard as hostile. At birth the Yanomamis' chances of surviving in their natural conditions was greater than that of people living in India, if we set aside the fact that these Amazonians practiced infanticide in the case of daughters.

A similar evolution, which has been called the "epidemiological transition,"[8] shows the role of the socioeconomic environment in the development of infectious diseases, which are, moreover, often replaced by cardiovascular conditions, diabetes, alcoholism, and criminality.

### *The immunity of populations and human activities*

Depending on their nature, human activities, even those that are a priori most praiseworthy, can expose populations to environmental dangers. We have already emphasized that farmers, in conquering untamed nature, have exposed populations to more or less serious illnesses. Before the destruction of the forests in West Africa, the *Anopheles gambiae* mosquito inoculated *Plasmodium falciparum* (the parasite that causes malaria) into pongids (chimpanzees, gorillas, orangutans, etc.); after the trees were cut down, the territory occupied by these apes receded, and the mosquitoes attacked farmers and their families. Poliomyelitis illustrates another form taken by the relationship between culture and infectious disease. Polio flourished in developed countries when modern methods of hygiene were applied. However, the illness is ancient; in the University of Pennsylvania archeological museum, a mummy dating from the thirty-seventh century B.C. shows the typical effects of this neurological disorder. Polio, a strictly human pathology, exploded in the first half of the twentieth century.[9] Improvements in hygiene, representing one way of modifying the environment, were not unconnected with this explosion. For example, in 1950, polio was much more frequent in Miami than in Cairo, while in Casablanca it struck young Europeans twenty times more often than it struck young Moroccans. These statistics might suggest that the ethnic difference between the victims was the determining factor, but in fact it was the difference in socioeconomic conditions that was crucial, because it affected the age at which children encountered the polio virus. European children, living under better hygienic conditions than the Moroccans, contracted the virus later, when they were no longer protected by maternal antibodies, which pass through the placenta or are carried by the colostrum, and got the disease with all its pathological signs. The young Moroccans, on the other hand, encountered the virus when they still had maternal antibodies, and this allowed them to take advantage of this passive immunity while their immune systems developed ad hoc defense processes. Under these conditions, the virus never found a "free area" propitious for the development of a serious infection.[10]

From 1906 to 1960, polio was an ongoing threat in the United States, illustrating the paradox of socioeconomic progress as a factor favoring the prevalence of a disease. American children, like the Europeans in Casablanca, encountered the virus at an age when they were no longer protected by maternal antibodies. Since 1960, the vaccine has allowed us gradually to control the epidemic, so that 99 percent of the cases worldwide are now practically limited to developing countries.

## THE IMMUNITY OF POPULATIONS EVOLVES

The cognitive capacities of the immune system evolve: the immunity of human populations is the result of a dynamic situation depending on the relationship, and in part on the power relationship, between microbes and the immunity of human groups. Humans' collective immunological heritage is the product of evolution and is probably always in evolutionary movement. No one any longer doubts that living beings have evolved;

the study of the processes involved in evolution, of their effects, and of the theories to which they have led is no longer limited to the traditional domain of biology but now extends to the social sciences and of course to ecology. Theories of evolution have been reviewed by Ruffié.[11] We will not discuss here the validity of each of these major theories; other writers have already done that. Let us simply note that nothing is completely settled so far as evolution is concerned: toward the end of the 1980s, a new hypothesis—that goes by the name "directed mutagenesis"—was proposed that challenged neo-Darwinian axioms. According to this hypothesis, bacteria can "choose" among the possible mutations the ones that give them the best selective advantage. This form of necessity without chance, which amounts almost to a molecular neo-Lamarckianism, was not unanimously accepted by specialists in genetics.

### *The immunological repertory: its gaps and its evolution*

The immune system of vertebrates has been evolving for six hundred million years. Each species has adapted to constraints represented essentially by microbes that have also evolved, the immune system and the microbes having coevolved. In fact, humans, microbes, and the environment have all three played an important role in this coevolution, and therefore the study of infectious diseases requires a truly ecological approach. The essential aspect of the evolution of the immune system is the acquisition of cognitive processes making it possible to recognize the self and the not-self. The structures participating in these processes are the receptors on lymphocytes. The antigen receptors on T and B lymphocytes probably evolved in accord with the neo-Darwinian model, with the persistence of the genes conferring a good ability to respond to microbes. However, it is not impossible that the lymphocytes capable of directed mutagenesis were able to choose, selecting the genes best adapted to their function of recognizing the not-self. The case is different for HLA molecules, because any given individual possesses only a limited variety of these molecules. The product of each HLA allele allows the presentation of numerous antigens, and, in fact, we do not know the maximum number of different antigens that a single allele can present. The HLA system can be thought of as the representative of the human race's immunological repertory. It is within a population that the federation of each individual's HLA repertory constitutes the super-repertory of the human group concerned. The HLA products of the population form both the community's self and the population's repertory. The more polymorphic the community self, the richer the repertory. The addition of HLA molecules—the structures for presentation of antigens—gives the population a greater capacity for immune response. The greater a population's HLA polymorphism, the better its eventual capacity for response will be. If the polymorphism has been able to evolve and, depending on the circumstances, vary in accord with a genetic drift or a founder effect, selection by infectious agents has probably played an important role in this phenomenon, insofar as better polymorphism leads to a better capacity for response. Natural selection would thus explain many associations between genes at the level of the MHC. When a human group does not have the HLA molecules that allow it to present antigens, these shortcomings in the immune response of the super-repertory represent a gap. We know that Amerindians who have a monomorphic HLA system, and hence probably a deficient repertory, have trouble immunizing themselves against vaccines. Within the human species, the HLA system has probably evolved under the influence of selective pressures exercised by microbes—but also in accord with other circumstances. For Jan Klein, one of the founders of modern immunogenetics, the HLA system, that "sphinx of immunology," underwent a trans-species evolution, the HLA alleles having already existed among primates in an ancestral form that was acquired at the outset by hominids, including *Homo sapiens*. Klein maintains that the evolution of the HLA system was slow, and not necessarily connected with microbes, although, as we have already mentioned, certain infectious diseases have been observed to be positively or negatively connected or associated with HLA alleles.

Moreover, animal species such as hamsters and rats have an extremely monomorphic MHC, which does not prevent them from battling infectious agents and adapting to the most diverse ecological niches on our planet. We have to recognize that this biological peculiarity is hard to explain if the main function of the MHC is seen as providing mechanisms of defense against microbes.[12]

The processes that generate polymorphism are probably neither unique nor exclusive; the presentation of antigens is certainly an important factor in evolution. We can suppose that under certain circumstances epidemics have led to powerful selective effects within human populations, "eliminating" individuals who were weak responders, and favoring the persistence of carriers of the good formula for defense against the microbes concerned. Microorganisms would thus have selected pre-adapted individuals, but they are probably not solely responsible for the evolution of the MHC.

If the polymorphism of HLA molecules is connected with selective pressures exercised by microbes, how is it that we do not see a larger number of solid associations between polymorphism and pathogenic agents? Even the most eminent experts on the MHC who met in Cambridge in July 1993 were unable to answer this question. It is easy to understand how, given the presentive functions of HLA molecules, the evolution and selection to which microorganisms led have participated in a more or less direct way in changes in the polymorphism of the MHC within communities.[13] It is just as easy to interpret the relationship between the resistance of the blood type known as the Duffy group to the malaria-causing *Plasmodium vivax*. Interpreting the evolution of the ABO blood system is more complex; in fact, setting aside situations in which the distribution of each blood group in the system has a historical cause, and taking into account the possible role of incompatible pregnancies tending to favor individuals with O-type blood, there are cases in which associations with pathogenic agents have not been satisfactorily explained.[14] The English serologist A.E. Mourant referred to a possible antigenic relationship between blood groups and infectious agents, a phenomenon that has still not been demonstrated. Associations between ABO groups and infectious diseases are usually weak, and their existence has in many cases been proposed on the basis of poorly documented investigations.

Nevertheless, even in the case of more recent investigations, such as those proving the association between the O group and cholera, we do not know the real reason or reasons for this phenomenon. Subjects who secrete into their saliva the substance of the ABO group are better protected against certain infections than are the nonsecretors, particularly in the case of urinary infections. In fact, protection follows the "drainage" of the bacteria after they have adhered to the molecule characteristic of the blood type, whatever it is. There is therefore no association with any particular blood group. This suggests that in the case of an association between a biological marker and vulnerability to a microbe everything involved is not strictly immunological in nature, and it is not impossible that the relationship of ABO to microbe or HLA to microbe does not always have an immunological substratum. Haldane was one of the first to mention the participation of diseases and pathogenic agents in evolutionary processes, and there is often a coevolution of pathogenic agents and their victims.

Myxomatosis in Australia is a good example of the joint evolution of host and microbe. It shows how rabbits with reduced vulnerability survived, and then how the virus evolved.[15] Jules Bordet, a great follower of Pasteur, pointed this out as early as 1947. It would be impossible to better describe the evolutionary influence exercised on humans by a pathogenic agent—for example, in the case of plague but also in that of other microbes that have affected human development almost from the outset. Malaria was and potentially remains an important selective factor in the presence of certain hemoglobins and blood groups. Trypanosomiasis, or sleeping sickness, which we have not yet mentioned, very probably played a role in the evolution of hominids. A very serious illness, it has raged in Africa since time immemorial and is due essentially to two parasites, *Trypanosoma gambiense* and *T. rhodesiense,* both related to a third, nonpathogenic microorganism, *T. brucei.* The tsetse fly transports the parasite to humans, and this vector, represented in the form of thirty different species, has existed for thirty-three million years. The fly and its infectious cargo certainly lived alongside the earliest African humanoids, exercising so much influence that for the evolutionary biologist and anthropologist Josef Reichholf they constituted the determining factor in humanity's migration out of its birthplace in Africa and its subsequent spread over the whole earth. It is now generally agreed that the host and the pathogenic agent coevolve, the whole being harmonized in a stable genetic equilibrium maintained by natural selection, as in the case of hemoglobin S and HLA alleles. Theoretically, natural selection, by producing resistant hosts, should lead to the appearance of microbial agents of increased virulence, but this is not usually the case, as we have seen in connection with myxomatosis. If we put ourselves in the position of the microbe, we can understand that a fatal disease restricts its transmission and persistence by eliminating the terrain that it infects. Thus, coevolution often produces less pathogenic agents that are adapted to their hosts in order to circumvent the defenses to the point that they "imitate," for example, tissue characteristics, that is, that microbes come to resemble, from the antigenic point of view, the constituents of human tissues and to produce "crossed reactions" with them.[16] Here the pathogenic agent adapts to the host without destroying it.

Over the centuries, pathogens may have forced man to pay a high price for his survival. However, thanks to his heritage of different genes for resistance to disease, man has forced microbes to adapt as well. Thus, a dynamic situation has been established between, on one hand, the host's resistance to microbes, and on the other hand, the efforts of microorganisms to circumvent this resistance by adapting, so far as possible, to the new ecological niche that a resistant host represents for them.

The flu virus, for instance, is a kind of champion in this area, to the point that the flu viruses of avian origin that affect humans are capable not only of evolving very rapidly but also of drawing on genetic information from their avian flu virus ancestors (this is called genetic rearrangement) in order to be more competitive in one way or another.

We mentioned earlier the important role played by antibiotics in the selection of infectious agents—an artificial selection that is involuntarily provoked but extremely important. Here we see the third "cultural" ("nurture") partner in the factors of coevolution, the other factors, the host and the microbe, being biological in essence. The case of Vietnamese living in the delta or the mountains has shown how culture and customs can also have an influence on the human-microbe relationship.

### *The immunity of populations, the immunity of all*

The preceding pages have explained how human groups have suffered and still suffer from the deleterious effects of infectious diseases. The various microbial factors ravage human groups by altering the adaptive immunological performances of a more or less sizable portion of the community. Those who are originally vulnerable or resistant to the harmful agents establish the vulnerability or resistance of the population. When the pressure of microbes is exercised during the period when the individuals concerned are capable of reproduction, it can diminish the performances of a vulnerable population by altering that population's capacities for reproduction, survival, and adaptation to the environment. Under the preceding circumstances, we can easily understand how microbes affect the group through its immunity. The phenomenon is less clear when observing older subjects whose reproductive years are over and whose potential for adaptive performance is in evolutionary theory without value for the community. In fact, we think that it is not in conformity with reality to dissociate older individuals from the rest of human groups with regard to the immunity of populations. Elderly individuals certainly participate, in a substantial way, in the altruism of the immunity of populations. Taking into account their immunity-related experience, they encounter epidemics as "naturally vaccinated" individuals capable of producing a notable antimicrobial response. The participation of the elderly in the conflict with microorganisms takes place at the price of a more or less elevated rate of loss among the aged people in the community, this way of "cutting losses" due to microbial diseases permitting the younger individuals to acquire their specific immunity. The case of women as a particular group with unusually effective immunity should also be noted.

### *The victory of polymorphism*

The heterogeneity of human groups—made up of ethnic variations, individuals of different ages, adults, the elderly, and women—contributes to the immunity of populations in diverse ways. Group immunity emerges from a collective and is produced by aggregating the immunological repertories of all individuals in the group, together with their respective competencies and immunological capacities. The HLA system is the molecular basis for both individual identity and the immunological repertory of the species. An individual's HLA system represents a portion of the population's immune potential, whereas the aggregate symbolizes the immunity of the population in its biological aspect, and this immunity enables the community to confront various attacks by other elements in the ecological niche (microbes, pollutants, and so on). The HLA system can be seen as the population's first immunological shield provided by human communities. It provides a good illustration of the contribution of human polymorphism to the processes of protection in different societies. HLA polymorphism is not the only factor involved in the effectiveness of the collective immunity in which individuals participate, in view of the role played by their immunological status at the moment, their ages their sexes their various systems of blood groups, and probably other biological systems that are not (yet) known to be factors in the immunity of populations. The genes involved in defense systems are the products of an evolution; in the case of the MHC, we see the influence of the choice of sexual partner, exogamy (a result of the incest taboo), and infectious diseases. Moreover, in a certain number of cases there has been a coevolution of both the host's processes of defense and the pathogenic agents.

When the host cannot adapt, or if the price of this adaptation is too high to pay, the host can still defend itself by avoidance. Take, for example, the case of birds in Hawaii. The archipelago, formed by more than a hundred islands, was a Polynesian colony for ten or fifteen centuries; Cook "discovered" it in 1778 and named it the Sandwich Islands. The European intrusion led to a near-disappearance of the indigenous birds, whose nests on the ground were destroyed by agricultural practices and the arrival of rats. The introduction of birds foreign to the archipelago was accompanied by the arrival of avian malaria (each living being is a kind of microbial zoo) whose agent, *Plasmodium relictum capistranoae,* is transmitted by the mosquito *Culex quintefasciatus.* The exotic birds that were resistant to avian malaria and plasmodium were already adapted to each other, whereas the indigenous birds, which were vulnerable to malaria, had to reduce their ecological niche to altitudes above fifteen hundred feet, a ceiling beyond which the mosquitoes could not fly. Thus, in Hawaii, the genetic polymorphism of birds, in terms of resistance to disease, takes the form of differences in the habitat chosen by the animals. Birds vulnerable to malaria have a limited ecological niche.

According to Reichholf, the first humans emigrated out of Africa for similar reasons. For him, the cause of this ancestral diaspora was the trypanosoma, the agent of sleeping sickness. As we have seen, the future Amerindians who crossed the Bering Straits on foot were perhaps fleeing a microbial calamity. The fundamental difference between humans and Hawaiian birds in this regard has to do with humans' knowledge, their culture. The birds do not know that building their nests at too low an altitude will rapidly expose them to disease; if they wander into the territory ravaged by the mosquitoes, then they are eliminated. When humans discover that a territory is too dangerous, they learn to recognize it and transmit what they have learned. When a group's immunological repertory does not allow it to adapt to a milieu, the practice of avoidance will become a cultural phenomenon. Thus for Reichholf the ancestral migration out of Africa was represented figuratively by the flight from Eden.

On the genetic level, the immunity of human populations has undergone selective, evolutionary processes not unlike those observed in other species. Nonetheless, humans are the only species in which culture participates in evolution; humans use their thought, their brains, in order to adapt.

### THE PERMANENT THREAT MODULATED BY HUMAN ACTIVITIES

Human societies have always tried, through their culture—through the religious, humanist, and scientific culture peculiar to them—to protect themselves from infectious agents by using an impressive panoply of real or supposed curative and preventive hygienic measures, including various medicines, vaccinations, and the like. A list of the methods that have been employed with greater or lesser success in the course of the history of humanity would be long and diverse, sometimes comical or sinister, ranging from useless "bleeding" to the prodigious antismallpox vaccine, and including sorcery, pogroms, the torture of lepers, and various religious practices. Humans have stubbornly battled infectious diseases, using procedures that have been thought up, tested, evaluated in accord with the most various criteria, and then transmitted through time. Human activities have sometimes fostered the development of microbial diseases. Progress such as the improvement of hygiene may have favored the emergence of a disease like polio, which was later eradicated by vaccination. Humans now possess more than ever the ability to be the yin and the yang of their relationship with microbes, being just as likely to facilitate the latter's proliferation as to attenuate their deleterious capacities. It is a notorious fact that wars, by bringing together thousands of individuals, creating disorder and famine and rending the social fabric, have favored the arrival of dramatic epidemics. It is generally agreed that Napoleon's defeat in Russia was due to his enemy's army and its ally, "General Winter," but that is to unduly neglect the role played by General Typhus, the main destroyer of the invading French army. Since the conflicts in ancient Egypt, Greece, and the Roman Empire, it has been known that states of war destroy the balance of the immunity of populations and favor microbial calamities. Typhus, plague, cholera, and smallpox have accompanied the combatants and contaminated populations. The most dramatic form of this kind of situation is represented by the conquest of the American continents and the disaster that followed. It is not surprising that military physicians in many countries have helped improve methods of treating and preventing infectious diseases; they did so simply with the goal of improving the effectiveness of their armies by protecting their own soldiers. After the massive mortality following the conquest of America, the First World War broke the record for mortality for an epidemic associated with a conflict. During that conflagration diseases were seen among combatants (typhus) and civilians (Spanish flu) on a previously unheard-of scale. During the Second World War, there was no such infectious cataclysm among soldiers who were vaccinated and better protected against typhus-carrying lice. In February 1944, for example, the epidemic of typhus that struck Naples was quickly wiped out by the use of the DDT, of which the American army had large stocks. In the course of this war, it was the deportees in concentration camps who represented, alas, an almost experimental model of the immunity of populations, showing the extent to which overpopulation, poor nutrition, stress, the absence of hygiene, excessive heat and cold, physical exhaustion, and physiological misery could sap the human organism's means of defense.

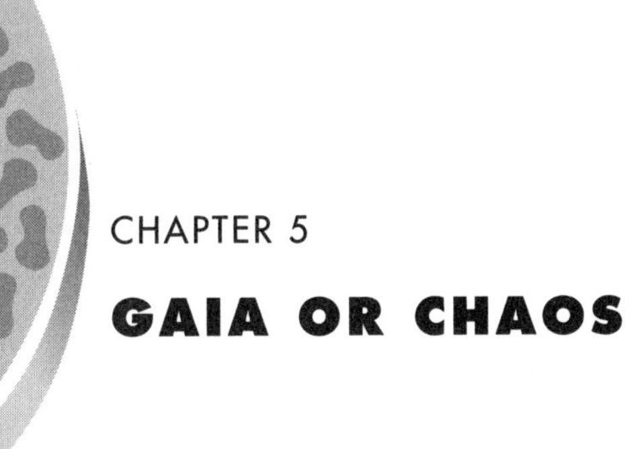

CHAPTER 5

# GAIA OR CHAOS

In Greek mythology, Gaia is the earth or, according to certain interpretations, the goddess of the earth. The Gaia hypothesis and chaos theory set forth two conceptions of the terrestrial world, and hence of the cosmos. The two ideas seem at first more or less opposed.

According to the Gaia hypothesis, proposed by James Lovelock, a British chemist, the Earth is a living being, the largest one we know. Chaos theory proceeds from much older work in mathematics, especially Henri Poincaré's. The term "chaos" itself is due to another mathematician, Jim Yorke, who popularized it.[1] According to James Gleick, a science journalist who has written about chaos theory for the general public, chaos theory was revealed to Yorke as he was reading a previously unnoticed article by a meteorologist at the Massachusetts Institute of Technology (MIT), Edward Lorenz. Lorenz's model was symbolized by the "butterfly effect," according to which the beating of a butterfly's wings in Brazil could lead to a tornado in Texas, this image describing the essence of chaos theory, "sensitive dependency on initial conditions"—about which I'll say more shortly. It is amusing to see the Gaia hypothesis and chaos theory, which describe two (apparently) different ecological processes, making use of symbols that are so bucolic and so complementary: daisies and butterflies.

## THE GAIA HYPOTHESIS

In chemistry, James Lovelock specializes in studies of atmospheric gases. In the 1970s he devised a gas chromatograph that made it possible to detect traces of a single gas within a mixture of gases. Because of his expertise, he was hired as a consultant by NASA and the Jet Propulsion Laboratory, where he worked on the possibility of extraterrestrial life on

Mars.[2] For Lovelock, it was incontestable that the composition of Earth's atmosphere (79 percent nitrogen, 21 percent oxygen, with traces of carbon dioxide, methane, and argon) was synonymous with life. The atmosphere of Mars, like that of Venus, with 96 percent carbon dioxide, 3–4 percent nitrogen, and traces of oxygen, was incompatible with any kind of life, or at least with any extended, generalized life on the surface of the planet. The Martian atmosphere is to Earth's what gases emerging from a car's exhaust pipe are to the air entering its engine. Earth is of cosmic origin, a product of the Big Bang, and it was on the basis of its specific constituents that life appeared on the planet at least three billion years ago when bacteria and algae capable of photosynthesis began to transform carbon dioxide into oxygen. In time, oxygen became a major component of the air we breathe. Consequently, all forms of life are emanations of the very substance of the world. There is no doubt that it was the emergence of life on Earth that made the planet what it now is. A central idea in the Gaia hypothesis is that life influences the environment. For Lovelock, Earth is a living being whose general homeostasis (temperature, oxidation, acidity, etc.) is maintained by feedback processes involving the biotope and biocenosis.[3] As we can see, Lovelock's conception is scientific-romantic; some people have called it a mythic notion of the Earth, in which life and the environment are part of one and the same system. It has been criticized as an almost supernatural form of order in which a superior director acts on regulatory systems in order to obtain a general harmony, as if species in biotopes cooperated with the goal of correcting, in the long term, deleterious processes. The Gaia hypothesis has, of course, provoked impassioned responses, both enthusiastic and critical. In order to show that the hypothesis does not require any kind of cooperation—no action based on prior agreement among living beings—Lovelock proposed his "Daisyworld" model.

### *Daisyworld*

Daisyworld is a planet that resembles ours in every respect except that it is populated only by white, black, and gray daisies. The temperature of this imaginary planet varies little or not at all, despite the sunlight. By their ability to reflect the sun's radiation (their albedo), the daisies influence the temperature at the planet's surface. The white daisies reflect solar radiation back into space, whereas the black ones capture it. Weak exposure to the sun's rays—the word for this is "insolation"—encourages the growth of dark-colored daisies, which absorb solar energy better, with the result that Daisyworld becomes a world covered with black daisies, and its temperature increases. Stronger insolation encourages the growth of pale daisies, whose multiplication and its effect on light reflection will bring about an overall decrease in the world's temperature, which had a tendency to increase, and so on. The controversy provoked by the Gaia hypothesis led to the organization of a symposium on the subject in 1988, and a book on it was subsequently published. In general, scientists have accepted the idea that the biotope could have an influence on certain aspects of the non-living world. They remain skeptical about other aspects of the hypothesis, however, in particular the idea of a Daisyworld that maintains its homeostasis not *by means of* the biosphere but *for* the biosphere. Although scientists had reservations regarding the Gaia hypothesis, the general public found it attractive, and, in particular, it enchanted those well educated in "New Age" philosophy.

It seems that the Gaia hypothesis grants little importance to the human race, which was, after all, a latecomer in the scenario of the biological world since its inception. According to Lovelock's conception, *Homo sapiens* is ultimately of little importance in a living world that is self-regulating and self-regulated. Thus, human agricultural, industrial, and cultural activities would have to be seen as elements that can be integrated into nature's feedback processes. The Gaia hypothesis has been criticized for conceiving the Earth as a living being that has undergone no overall mechanism of natural selection, a concept that is hard to accept in an evolutionist hypothesis. Daisyworld's supporters reply, citing mathematical proofs, that natural selection produces a localized improvement and an immediate advantage rather than a long-term benefit. According to them, natural selection would not always be a crucial factor in the evolution of certain systems. We know that "chance and necessity" cannot explain everything in evolution, because, as Kimura has shown, chance without necessity—that is, a neutralist theory of evolution—is also part of the truth and of scientific reality. (I explained Kimura's theory in note 11 to Chapter 4). It is clear that matters are pretty complex, and everything depends on how one perceives Gaia: is it an ecological model, a metaphor, a myth, or a philosophical movement? In any case, Gaia has had, and will continue to have, various kinds of success.

### *What are Gaia's virtues?*

Could Gaia correct all ecological aberrations? If so, we wouldn't need to be concerned about the future of the Earth. After all, it would be possible to do anything we wanted, and Gaia, through its self-regulating functions, would correct the aberrations and reestablish a balance. Thus Gaia would remain a living being, which does not mean that our beloved planet would always be able to support human life. Gaia was already a living being when there were only bacteria on Earth; in the event of an ecological catastrophe, might it not become once again the living being it once was? In other words, is the human species necessary to Gaia's survival?

Some people, more prudent, will prefer the conception offered by Lynn Margulis, who for many years was Lovelock's collaborator. In the book, *The Biota and Gaia,* Margulis writes:

> Rather than state "Earth is alive," a phrase that confuses many and offends others, we prefer to say that Gaia is a hypothesis about the planet Earth, its surface sediments, and its atmosphere. We describe the Gaia hypothesis as follows: the Earth's surface is anomalous with respect to its flanking planets, Mars and Venus. The surface conditions of Mars and Venus can be adequately comprehended by physics and chemistry. With respect to certain attributes, the Earth is, from the vantage of physics and chemistry alone, inexplicable. The Earth's physical and chemical anomalies, given new concrete knowledge about Mars and Venus, have become obvious. They include the presence of highly reactive gases (including oxygen, hydrogen, and methane) coexisting for long times in the atmosphere, the stability of the Earth's temperature (that is, the long-term presence of liquid water) in the face of increasing solar luminosity, and the relative alkalinity of the oceans. The pH of the Earth is anomalously high. When compared with its barren neighbors, Earth's surface chemistry is aberrant with respect to its reactive gases, its temperature, and its alkalinity. These discordant chemical and physical conditions have been maintained over geologic periods of time. Lovelock's concept, with which we entirely agree, is that the biota (that is, the sum of all the live organisms at any given time), interacting with the surface materials of the planet, maintains these particular anomalies of temperature, chemical composition, and alkalinity. Therefore, to understand the Earth's surface we must understand the biota and its properties; we can no longer rely only on physical sciences for a description of the planet."[4]

### CHAOS

In the 1960s, Edward Lorenz, the MIT meteorologist, created computer programs for use in climatological and meteorological studies.

The data entered in Lorenz's programs included the ambient temperature, atmospheric pressure, wind speed, humidity, and so forth. As he worked, Lorenz discovered that a tiny modification in one of the data entered in the program could have extremely important effects on the phenomena observed. He called this "sensitive dependency on initial conditions" and symbolized it by the metaphor of the "butterfly effect." Lorenz's basic idea is that there is always a degree of unpredictability in a complex dynamic system such as an area's climate. Even if we knew the state of each molecule in the atmosphere in question, there would always be something unknown in the system's development. In a nonlinear system an element that seems negligible can have extremely important consequences. Chaos thus marks a considerable departure from the ideal determinism proposed by Pierre-Simon Laplace (1749–1827), a French mathematician, astronomer, and physicist, in his *Analytic Theory of Probability*. Laplace maintained that if some demon knew the properties of each object in the universe at any given instant, he could predict the universe's future development.[5] As we have already mentioned, Henri Poincaré was a pioneer in the study of chaos.[6] Fractals were described by Benoit Mandelbrot, a Polish-born mathematician who emigrated to France in 1936.[7] Nature and the world of living beings include an infinite number of fractal structures. It is hard to imagine that DNA can contain information for the incredibly complex structure of, say, the bronchial network, from the trachea to the alveoli. It is likely that the genetic code must, as Gleick puts it, "set the rules of an iterative process of bifurcation and growth." On the basis of this genetically determined information a process governed by chance is supposed to lead to a finished product. In the world of the living, there exists a form of order proceeding from the chaos represented by certain organs and living creatures that are fractal (that is, some part of them reproduces the whole, as in the case of bracken and coral).

Chaos is not just a "mess," and it is not chance, either; rather, theorists speak of deterministic chaos. Chaotic systems are governed by a form of determinism; they are extremely sensitive to the initial conditions, and behind their apparent disorder there is order and a model. From the order that could have issued from chaos, periodic phenomena may appear during the evolution of complex systems. This has been demonstrated experimentally, as in the case of Liesegang's rings, Bredig's heart, and Zhabotinski's reaction. The examples cited belong to what Ilya Prigogine has called "dissipative structures," because they consume energy in order to maintain and develop themselves. Liesegang's rings are obtained by precipitating a solution of silver nitrate with a salt of another metal. In a thin layer of gelatin, concentric rings that form for several hours are observed. Bredig's heart is obtained by putting a drop of mercury at the bottom of a container filled with oxygenated water. The drop is oxidized and then loses its oxygen, then oxidizes again, and again loses its oxygen as it changes form (contraction and relaxation). The drop of mercury "beats" like a heart. Zhabotinski's reaction is the slow oxidization of malonic acid by cesium sulfate and potassium bromide. In a mixture of these reagents one can see the formation of colored, superimposed layers. These are self-organized structures with emergent properties such as the self-organized criticality described by Per Bak. Per Bak, a physicist at Brookhaven National Laboratory, thinks complex dynamic systems evolve toward a critical state (self-organized criticality). The experimental model he uses is a sandpile. By letting a thin stream of sand flow from one's hand, a conical pile is obtained on which, as the sand accumulates, small slides, and more rarely, large ones, occur. A pile susceptible to sliding represents the critical state of a system in equilibrium. An additional grain of sand can lead to a large avalanche (sensitive dependency on initial conditions).

Chaos, with its laws and its representations (fractals), does not exist solely in the domain of climatology, in that of biology, or only for those who, like Bak, like to play in the sandpile. Respected scientists have applied the model to astronomy (Poincaré) and also to physics, chemistry, the human sciences, economics, and even politics. Ilya Prigogine, a Belgian of Russian origin and winner of the Nobel Prize for Chemistry, has contributed to the general public's knowledge of chaos theory by publishing a remarkable work, *The New Alliance.* For Stuart Kauffman, one of the pillars of the Santa Fe Institute[8] and a physician and mathematician, we live in a world characterized by a general compromise between order and chaos, which he explains in his two books.[9] Kauffman also alludes to the Austrian logician Kurt Gödel's[10] principle of undecidability and to his incompleteness theorem in explaining the impossibility of predicting the future.[11]

## THE QUESTION: GAIA OR CHAOS?

Here, we contrast Gaia with chaos in order to discuss the general processes underlying the emergence of infectious diseases that can have disastrous consequences for human groups of varying sizes. We have little interest in comparing two models, one of which, Gaia, is still only a hypothesis often rejected by academia and by the scientific "establishment." As Roger Amos Lewin, an anthropologist and scientific writer, has indicated, the Gaia hypothesis has been criticized for its apparent or real notion of a determinism that expresses an intrinsic goal, a kind of teleology. In short, Gaia has been reproached for being only a pseudoscientific myth. For us, the main interest of Gaia is not the hypothesis that so often annoys professors (probably because it has been so successful with the public). Rather, Gaia is interesting as a metaphor symbolizing a harmony, a self-regulation, that may in fact be a little magical insofar as it makes of the planet Earth an organism with a marvelous, self-regulated homeostasis. Chaos theory, on the contrary, represents a model of a nonlinear dynamic system, and there are good reasons for thinking that as a theoretical representation it is fairly compatible with the state of fragile, permanent equilibrium existing among the higher living species and infectious microorganisms. Over the past two decades, an abundant literature applying chaos theory to the ecology of populations has appeared, using in particular the Lokta-Volterra predator-prey model.[12] The Volterra model applies to epidemics only with serious reservations, because equality a microorganism to a predator is in many respects erroneous. In a general review of this model, Sabin and Summers have demonstrated that the answers to the questions asked about epidemics could be obtained by using chaos theory, but there are good reasons for thinking, and examples prove it, that approaches to the relationship between pathogenic microorganisms and hosts are in conformity with the model of sensitive dependency on initial conditions.

## SEALS AND DOLPHINS: GAIA OR CHAOS?

Let us now extend our discussion to living species other than humans in order to emphasize that the problem of life amid microorganisms is not limited to *Homo sapiens* but affects other higher animals as well as plants. Among the species in question are marine mammals, species that inhabit the element that covers most of the globe.

In 1998, the mortality rates of seals and sea lions were still very high in the subarctic areas south of New Zealand. In 1997, the population of Hooker sea lions (*Neophoca hookeri,* one of the world's rarest species) in the region was estimated at about fifteen thousand. Over the preceding decade, New Zealand's Department of Conservation had kept watch over mating sites at the sea lion colonies on Dundas and Enderby islands, 420 kilometers south of the mainland. These two colonies represent 95 percent of the population of reproducing Hooker sea lions. About twenty-five hundred pups were born each year, 80 percent of them on

Dundas Island, which is more difficult for humans to access than its neighbor, Enderby Island. Biologists who visited Dundas in January 1998 found about seven hundred young animals that had recently died or were moribund, whereas the adults seemed healthy. Gradually, the females began to show signs of illness during gestation. In early February 1998, it was estimated that a minimum of 41 percent of the young and 20 percent of the females were dead. The population of adult animals on the beaches was much smaller than normal, suggesting that an unknown number of males and females perished at sea. The possibility of infectious disease seemed so serious that the New Zealand minister responsible for protecting the animals forbade tourists to visit the islands, in order to avoid transmission of the disease to humans and to prevent tourists from spreading the disease to uninfected animals.

The cause of the New Zealand sea lions' misfortune is still not known. Experts are unsure if it was caused by pollution or by infection, as in the case of the death of Lake Baikal seals (*Phoca sibirica*) in the late 1980s. In 1987, about 70 percent of the seal population on the lake had died, an extraordinary mortality that was at first attributed to pollution, given Soviet authorities' indifference to ecological considerations. In fact, the following year there was a high frequency of abortive pregnancies among *Phoca vitulina* seals. The year after that, hundreds of adult seals died in the North Sea and in the Baltic, from Sweden and Denmark to Scotland. It was quickly proven that the seals were dying from infection with a morbillivirus (a family of viruses that includes the measles virus). The animals on Lake Baikal were contaminated by a virus different from the one on the North Sea; in the USSR it was the PDV-2 virus (phocine distemper virus-2), practically identical to the virus that causes distemper in dogs. On the North Sea though, the infectious agent was a new virus that caused a new disease, a pathology connected with an unknown morbillivirus that has since been named phocine distemper virus-1 (PDV-1). The seals in the North Sea thus found themselves confronted by a real problem of the immunity of populations. In the presence of a new pathogen, these animals used their immune supersystem (the MHC molecules described in Chapter 4), the federation of the immunities of each individual, in order to grapple with the pathogen. Moreover, it is noteworthy that many animals showed signs of immune deficiencies. The reasons for this deficiency are unknown; were toxins or ecological factors (in a very broad sense) involved, or was the deficiency caused by the infection itself?

At about the same time, dolphins in the Mediterranean were reported to be dying from lung problems; they were found dead or moribund on the coasts of North Africa, Spain, and France. These dolphins, like the North Sea seals, showed troubles with their immune systems, to the point that people spoke of "dolphin AIDS." In fact, the cause was a morbillivirus that also infected porpoises and was closely related to the measles virus.[13] It turned out that these marine mammals had fallen victim to different viruses: PDV-1, which, like PDV-2 that attacked Russian seals, is similar to the virus that causes canine distemper, porpoise morbillivirus (PMV), and dolphin morbillivirus (DMV).[14] The epidemic among Lake Baikal seals is thought to have been transmitted directly from dogs after residents of the area threw into the lake the bodies of sled dogs that had died of the disease in 1986. The story of Lake Baikal illustrates a phenomenon well known in infectious pathology: the increased virulence that a microbial agent can acquire when it passes from one species to another, in this case from dogs to seals. But there are few explanations for other epidemics that affect sea animals in the various oceans (including the Atlantic) and are caused by viruses that are not among the classic ones. These epidemics are thought to result from the combination of climatic problems (cold winters), major pollution, in particular by pesticides,[15] and, of course, the infection itself. It is interesting to note that these various epidemics affect species of mammals living in a single milieu—the seas—but at different sites, and that the animals were struck by related but distinct viruses. For example, no clear relationship has been found between the virus that affects Mediterranean dolphins and the other known varieties of morbillivirus.[16]

## WHO IS TO BLAME? WHAT IS TO BLAME?

Fatal infections among marine mammals raise a real problem for anyone who wants to discover the cause of the diseases. To be sure, the direct causes are clear enough: viral infections, often involving the same group of viruses, whose reservoir and mode of transmission are known, as in the case of the epidemic on Lake Baikal. In other epidemics matters are more complex. The "globalization" of diseases suggests a global cause, a toxin, or other form of pollution. These events are all the more troubling because the animals manifest a more or less profound immune deficiency and because we do not know whether this deficiency precedes or follows the viral infection. One of the factors that may be involved is the alga *Ptychodiscus brevis*, which produces a powerful toxin affecting the nervous system. The proliferation of this alga has been attributed to the combined effects of climatic changes and abnormally high rainfall with the resulting runoff carrying into seas an abundance of nitrogen provided by human and animal excrement. However, this hypothesis explains neither the observed immune deficiency nor the origin of the virus involved, even though the viruses could have originated on land.[17]

What is true of the emergence of human epidemics is also true of animal epidemics: the phenomenon is complex and a germ alone cannot make an epidemic, far from it. The modifications of our environment favor the appearance of epidemics, and the case of marine mammals shows how important this ecological aspect is: animals living in the same milieu almost simultaneously develop infections due to viruses that are related but nonetheless different in important ways. It seems that as a result of biological (algae) or abiotic (synthetic chemical substances) attacks on their environment, marine mammals are "offered up" to the most virulent pathogen or possibly to several dangerous microbes.

Are we living in a world like that of the fourteenth century when the Black Death struck—a world whose natural system is enslaved, a world of white and black daisies—or are we in a situation that has been unfortunately provoked by the beating of some nefarious butterfly's wings?

### *Theory and reality*

Recent theoretical and practical approaches to the demography of epidemics tend to study measles epidemics. From the end of World War II until the advent of vaccination, these were the epidemics for which we had the best data about incidence and distribution, at least for western Europe, North America, Australia, and New Zealand. Although there is a difference between theory and observation, the chaos hypothesis is now widely accepted by researchers in this field, in particular because it explains the unpredictability of the phenomena involved.

In reality, it is not possible to account for an epidemic's development without taking into consideration the immunological state of the population concerned, the so-called "herd immunity,"[18] though a herd, of course, lacks the cultural aspect of what we call the immunity of populations. Leigh Van Valen has proposed a hypothesis that suggests, as Claude Combes has put it, that "the most important component of our environment for any given living species is represented by the other species with which it interacts in the ecosystem."[19] In other words, the host and the pathogen coevolve. "Coevolution is the process by which two adversaries constantly acquire new adaptations in order not to be outrun by the 'other.' There is a series of mutual selective pressures." Coevolution is observed in the case of simian immunodeficiency virus ($SIV_{mac}$). Sooty mangabeys have a high rate of viruses circulating[20] without immune deficiency. The absence of any pathological manifestation of the virus is not due to a particularly effective defense mechanism, but rather to the mangabey's adaptation, which lessens its susceptibility to viral infection. Modifications of the CCR5 receptor have made the mangabey's immune cells "impermeable" to $SIV_{mac}$.[21] Indications are that both the virus and the mangabey's receptors have changed the relationship of susceptibility between the microbe and its host.

The coevolution of host and microbe is among the richest of all subjects of intellectual speculation in the science of infections. Just as Alice and the Red Queen move together with the environment, so the microbe and its host, the aggressor and the victim, make use of compensatory adaptations over time. Usually the microbe, by inventing new forms of virulence, "draws" its host along, but this is not always the case. Let us simply say that, as a rule, the microbe generates one or more mutations that are harmful to the host, and the latter, by evolutionary feedback, sets up a defense mechanism adapted to the microbe's new form. This kind of schema makes sense only in the context of chronic conflicts or long-lasting pandemics (plague, cholera, malaria, AIDS, for example.). However, the coevolutionary model cannot be applied to human beings without taking into account the "cultural" aspects of our species' relationship with microbes. Mathematical models of epidemics must also include the cultural aspect of the coevolution of humans and microbes.[22] Interpretation is still more difficult when the agent responsible for an epidemic makes use of an intermediary host. In the case of malaria, for example, coevolution can potentially contribute to modifications of the genes of humans, the parasite, and the mosquito involved. As one writer in *Science* observed, "It has been shown that malaria has evolved to transform the mosquito into the equivalent of a Count Dracula thirsting for blood."[23] In short, there is currently general agreement regarding the existence of a coevolution between the host and the pathogen.

Group immunity, particularly in the case of human groups, depends on the provision of maternal immunity (for newborns),[24] vaccination; the infectious risk itself; the nature and the frequency of contacts with the pathogen; and the biological, social, and cultural problems that are involved in the immunity of populations. Let us also mention the importance of vaccination, which has transformed the problem of the immunity of human groups with respect to smallpox, just as it was later to transform it for polio. Retrospective study of epidemics of smallpox in England has shown how difficult it was to formulate a single theoretical model. Various authors have used parish death registers as well as the information collected in "bills of mortality," and have thus been able to see that the progress of epidemics differed, depending on the period (seventeenth and eighteenth centuries), the groups concerned, and the size of the communities (small towns, middle-sized town, large cities).

Regarding the choice between a linear model and nonlinear model that includes the causes of phenomena (sensitive dependency on initial conditions), the question of the validity of mathematical modeling in general can be raised. Can a single model be employed in the case of epidemics, especially when human groups are affected? In fact, there is no reason not to apply one or more biomathematical models to a phenomenon that involves simultaneously demography, predator-prey relationships, epidemiology, and so forth. Among these models, chaos theory is incontestably the best-adapted to the study of epidemics, and the misfortunes of the marine mammals described above are in accord with this hypothesis. If so many animals of different species are affected by clearly distinct viruses, that is more indicative of germs operating in a situation favorable to their proliferation than of a phenomenon whose prime mover is the viral emergence itself. Sensitive dependency on initial conditions, probably being the same for all these marine mammals, fits with the globalization of an event involving many factors. As we noted earlier, in the case of epidemics it is counterintuitive to accept a single theoretical model. This is already true in the case of a rather sudden and rapid microbial attack, and it is still more difficult to bring under a single model entirely different events such as are involved in most diseases caused by parasites. In the latter case, we are confronted, as Combes wrote in his book *Interactions durables,* with a world in which the parasite's genome has a permanent adaptive influence on the host's genome and vice versa. Here we find, in a basal state, a relatively unstable equilibrium, a kind of biological "armed peace" that may last for ages, as in the case of malaria or sickle-cell anemia. Individuals who are chronic carriers of parasitoses are not like healthy carriers of infectious viral or bacterial diseases. In viral infection, aggressor and victim have a biological relationship, whereas bacterial disease carriers serve as nests of quiescent microbes without there being any true infectious relation between microbe and carrier. For example, some individuals carry meningococcus, the bacterium that causes meningitis, without any pathological effects, and without an asymptomatic form of the illness, either. In the case of long-lasting interactions of a parasitical nature, the host constantly makes use of its immune processes, while the parasite constantly deploys its aggressive instruments. Nonetheless, as Combes notes, " . . . these equilibria are extremely fragile. Just as in an individual, if malnutrition or any other cause leads to a weakening of immunity, the parasitosis can immediately become serious, so in a population, if the climatic conditions, for example, increase the success of transmission, parasitosis can become devastating."[25]

As was mentioned in Chapter 4, the relationship between immunity and microbes is not a conflictual one leading, according to a rather simplistic model, to the triumph of one of the antagonists. When it cannot win a definitive victory (as in the case of long-lasting interactions), the immune system can often adapt to this dangerous proximity. It is as though the immune system adopted a strategy inspired by Taoist philosophy. Modeling a reality or rather realities as complex as epidemics amounts, of course, to an ultrasimplification. It involves putting on the blackboard the factors, usually small in number and often very simple (number of contaminated individuals, number of deaths, and so forth) and then applying more or less arbitrary rules to them. Thus it is a matter of using a procedure that is related to the structural method as defined by the French linguist and philosopher Abraham Moles. With epidemics, we are clearly in the "domain of sciences which, in the current state of things, are imprecise, and will remain so for a long time."[26]

Without denying the interest of theoretical models in the study of events as dynamic and complex as emergences of infectious diseases, it seems to me that in this case theory stumbles: it encounters difficulties in providing an exhaustive explanation of complexity. These difficulties have been very well discussed by Jean-Pierre Dupuy in his *Enquête sur un nouveau paradigme*.[27] But do we really need to choose between the idealized vision of the planet Gaia—which is supposed to soften every ecological blow, including those produced by the conflict between *Homo sapiens* and microbes (the latter being, let us recall, the Earth's first inhabitants)—and an approach that refers to the ultimate stage, the science of chaos, the macroscopic emergence of cellular processes and dissipative molecules? The Gaia hypothesis, which is reductionist, treats the world as a field of daisies (or their equivalents), forgetting that if life has transformed the planet, humans have just as often domesticated the expression of life, as their technology has created a second nature. When we refer to Gaia, we should keep in mind the anthropomorphic aspect of the model: this Gaia, which has been modified and modulated by human occupation (and by occupation we mean activity and possession), cannot, of course, be reduced to a living world limited to its biological and physical/chemical aspects alone. Humans have transformed Gaia into a meta-Gaia, a metaphor designating a complex real world. While contemporary science may seek, in accord with its mission, to help explain nature, if we have to choose between a deterministic view of the world and one that delivers it over wholly to chance, reference to deterministic chaos seems the wisest option. In summary, I propose to replace the choice between Gaia and chaos by an aphorism: MetaGaia *and* chaos.

## MICROBES CAUSE MORE THAN INFECTIONS

The species *Homo sapiens* will have to make choices in order to protect itself against harmful infectious agents, but before examining the problems raised by this project, it will be useful to emphasize that the threat posed to human health by microbial agents, and especially by viruses, is not limited to acute or chronic infections. It is now certain that there are diseases known as "autoimmune" and cancers that result from viral or bacterial infections. That is, microbial attacks lead not only to an infectious danger that we might consider conventional, but also to disturbances that can even put an indelible stamp on the genomes of the individuals affected. The MetaGaia and chaos model applies not only to epidemics, to more or less resounding infectious outbreaks, but also to more torpid, less obvious forms of contamination by microorganisms that constitute, so to speak, biological time bombs.

### *Cancer*

Cancer is a disease that has been known for a very long time. The word "cancer" comes from the Latin for "crab" and was used as early as the fifth century B.C. While the important role played by DNA and its anomalies in the occurrence of cancers is generally recognized, environmental factors are also thought to play a determining role. These factors are said to be involved in 70 percent to 90 percent of all human cancers. That does not mean that external factors cause most cancers but only that they contribute in an important way to the occurrence of tumors and leukemia. We know that among these agents are viruses and at least one bacterium: *Helicobacter pylori*.

In 1910, the American biologist Peyton Rous was working in a laboratory at New York's Rockefeller Institute. He noticed that he could cause a sarcoma, one form of cancer, in a chicken by injecting it with ultrafiltered extracts—that is, viruses—taken from the sarcoma of another chicken. The virus was named Rous sarcoma virus, but his discovery, which was regarded as an exception in the animal world, was soon forgotten. In the 1950s, Ludwig Gross discovered a virus capable of provoking leukemia in mice; it now bears his name. Rous and Gross are two pioneers who showed the infectious nature of cancers in animals. The idea that human cancers could also have a microbial origin is connected with the work of a British surgeon, Denis Burkitt, who practiced in Uganda. Burkitt had noticed the high frequency of jaw tumors in Ugandan children who lived at altitudes below 3,700 feet, whereas children living at higher altitudes did not develop these tumors. The epidemiological data suggested that malaria was involved in the occurrence of this cancer, which was henceforth known as Burkitt's lymphoma. In the 1960s, the role played by Epstein-Barr virus in these cancers was demonstrated. The virus, transmitted by saliva, causes infectious mononucleosis in Europe and North America; in China and North Africa, it causes a cancer of the nasopharynx. Here is a virus that, depending on the established ecological or ethnic conditions, provokes very different diseases ranging from common infectious mononucleosis to the serious Burkitt's lymphoma. Why are there such differences? What factor or factors are determinants in the advance of the pathology? The genetic heritage of the individuals affected? The ecological niche? Both of the preceding?[28] It is not my purpose here to determine the causes of cancer, but rather to emphasize that microorganisms, which constitute the great majority of Gaia's living species, are sometimes involved in the occurrence of malignant tumors. This idea is not new, because as early as 1773 the French physician Bernard Peryilhe suggested that cancer was "viral" in origin, a hypothesis taken up again by Amédée Borrel at the beginning of the twentieth century. Table 2, which draws on the German oncologist Harald Zur Hausen, shows the principal viruses participating in the occurrence of human tumors. In this context, they are called oncogenic viruses, even though they are usually responsible for infectious diseases that are fairly benign. At present, many reports implicate viruses in the occurrence of other cancers, such as Kaposi's sarcoma, so frequent in the course of AIDS, and myeloma, a tumor of cells in the immune system. In both cases a herpesvirus is suspected.

As we have seen with regard to Burkitt's lymphoma, the infectious agent is not the only factor in the occurrence of the cancer; environmental, individual, and genetic factors are also often involved. Researchers have been

TABLE 2 THE PRINCIPAL VIRUSES RESPONSIBLE FOR THE OCCURRENCE OF CANCERS IN HUMANS

| *Virus* | *Infections* | *Cancer* |
|---|---|---|
| Epstein-Barr | Mononucleosis | Burkitt's lymphoma |
| | | Cancer of nasopharynx |
| Hepatitis B virus | Hepatitis B | Liver cancer |
| Papillomavirus | Papillomas | Skin cancers |
| | Warts | Genital cancers |
| Leukemia T virus | Paralysis | Leukemia |

concerned about the possibility that certain human groups might be more susceptible to the generation of tumors than others. The use of the usual biological markers reveals associations between ABO blood groups and HLA characters, on one hand, and malignant tumors on the other. Nonetheless, we have to recognize that none of the reports is wholly convincing, and that "group predisposition" (as opposed to familial predisposition) is not a scientific domain that interests the scientific community.

### *Autoimmune diseases*

In autoimmune diseases, the immune system is deflected from its normal function, namely, self-protection. MacFarlane Burnet, a pioneer of modern immunology, was the first to observe the intimate processes of an individual's self-destruction by his own immune system. A person who suffers from an autoimmune disorder is a victim of an attack carried out by the cells and antibodies of his own immune system heritage. An autoimmune disorder results from the disturbance of a potentially unstable equilibrium, the rupture of a harmony between the physiological and the pathological. This kind of rupture can be observed in the case of aging, as if aging were a sort of autoimmune disorder. It is generally acknowledged that the occurrence of an autoimmune disorder results from the conjunction of environmental processes and genetic predisposition. The association of these two elements (and perhaps others) elicits an auto-aggressive process whose main characteristic is a chronic inflammation. This prolonged inflammation is very harmful to the patient. Among the factors of the microenvironment, infectious agents are implicated, as, for example, in the case of arthritis of the vertebrae, a kind of inflammatory rheumatism that is very painful and crippling and affects one person in a thousand. The symptoms of this disease are caused by an attack made by the patient's own immune cells, but they are provoked by infection by *Klebsiella, Salmonella, Shigella, and Yersinia* bacilli. Patients suffering from this rheumatism experience symptoms that last for years after the infection. It is remarkable that 90 percent of these patients carry in their tissues the genetic trait HLA-B27, which is present in only 9 percent of individuals in good health. This is a pathological situation that associates the microbial environment with an individual genetic predisposition.

### *A time for choices*

Antoni Van Leeuwenhoek is known as the inventor of the microscope, and it is to him that we owe the first observations of bacteria. This erudite polymath was also interested in human demography. A Dutch citizen who realized that the surface of the habitable earth was 13,385 times larger than that of his own country, and taking as a basis the population density of Holland, he calculated that in 1679 the world's population was 13.4 billion. This overestimate merely anticipated later developments: in 2050, there will be seven to eleven billion people on Earth.[29] Human population is increasing, and the distribution by age groups is changing. Since 2000, there have been more old people than young. Moreover, today most women have, on average, two children, that is, scarcely enough to maintain the current population. Finally, the distribution of people between urban and rural areas is changing; in 2007, the number of people living in urban areas will be greater than those living in rural areas, and among those in cities, an extraordinary percentage will be living in slums. In short, human beings are increasingly numerous, and schematically, increasingly older and living in cities where living conditions are often marginal. Thus imagining the future in terms of the immunity of populations comes down to reflecting on the immunity of populations of "elderly" people.

Older men and women constitute increasingly large communities in industrial countries, and for a student of the immunology of populations they represent a specific group that is potentially more vulnerable to the various dangers faced by the immune system.[30] According to current paradigms, an elderly person's immune apparatus is mediocre in quality. Any additional attack on this deficient apparatus is likely to have serious consequences for the individual involved, since his declining immune system has trouble fending it off.[31] This was shown—in a nondogmatic way, of course—by *in vivo* and *ex vivo* investigations that confirmed the existence of the elderly individual's immune deficiency. For example, when old people are vaccinated, the number of them who are capable of producing antibodies decreases steadily after the age of sixty, and the quantity of antibodies produced by those who are still capable of doing so also decreases.[32] Immunological senescence thus takes the form of deficiency. The environment can have devastating effects on elderly persons' defense mechanisms; we know how serious hospital-acquired infections, flu epidemics,[33] and bacterial pneumonia[34] can be for them. In addition, evaluating the immune response of elderly individuals can help us predict their life expectancy.[35]

Nonetheless, let us not restrict our discussion to what is now a cliché. First of all, we have to define what we mean by aging. For the biologist, the term designates anatomical and functional changes that accompany the individual's maturation.[36] In this sense, aging begins with birth. For the gerontologist, aging corresponds to anatomical and functional changes connected with age and occurring after sexual maturity. We believe, with Ruffié, that there is "no well-defined borderline between the adult and the elderly."[37] From puberty on, the immune system shows a shrinking of the thymus connected with age;[38] if we see that as a manifestation of the aging of the immune system, then this aging begins early and it is programmed by a large number of genes. We can, moreover, ask to what extent environmental constraints, beyond genetic programming and during the life of the adolescent and adult, can compromise the quality of the aging person's immune system, and what effect(s) this might eventually have on the immunity of populations. A student of the latter subject must, in order to understand the problems raised by the immunity of the elderly individual, know what events affected his processes of maturation when he was young.

However, aging does not depend solely upon the environment; an important part of this phenomenon is genetically programmed.[39] About two hundred genes are estimated to be involved in the evolution of *Homo sapiens*' life expectancy.[40] The performance of the immune apparatus follows the neo-Darwinian model of natural selection, according to which within a population under strong selective pressure connected with the environment (an epidemic, for example), only those individuals will survive who have an immunity that allows them to eliminate the pathogen(s). A priori, the relationship between immunity and aging does not seem relevant to the question of a possible natural selection, insofar as it is not clear how individuals could, beyond the period in which they are fertile, propagate advantageous genes. However, if we admit that the elderly were once young people whose immune system allows adaptations, then we can suppose that before the advances in medical science we have discussed, the elderly represented, and may still represent in countries with deficient health care systems, the "products" of a selection. That is the case in developing countries, and in developed countries, it is the case for persons born in the early twentieth century, before "cultural" or medical adaptation was added to biological adaptation. It is not easy to find models that enable us to gauge how environmental influences on the young might affect the process of aging. So far as infectious diseases are concerned, the histories of epidemics give us minimal information. At most, we can say that survivors of epidemics had a good immune apparatus, and, in short, that getting old was not something everyone could do, even if human groups were more or less exposed depending on their sociocultural milieu.

Without neglecting the inadequacies of elderly people's defense mechanisms, it is more correct to say that in the majority of human societies, the group immunity of elderly individuals is operational, effective, and allows a genuine response against microbes. This response suffers above all from not being sustained because of these persons' general physiological state, which does not permit them to mount the kind of prolonged defense that fighting off a pathogen often demands. On the other hand, the advantage of the elderly comes from their immunological past. Their immune system is worn down, to be sure, but thanks to lymphocytes, it has memorized a defensive know-how that allows elderly individuals to present means of protection, those of the specific immune response, that are already effective in the initial stage of the infection. Even those who cannot long endure the deficiency of their general condition when aggravated by microbial infection still participate in the group defense and in this case constitute the group's way of "cutting its losses" in an epidemic. Thus in humans aging is not an unfavorable factor for the immunity of populations, despite the problems sometimes raised on the social level by the potential disadvantages of the aging of populations. Let us recall that the Spanish flu was particularly hard on young men, much more than on women and, in the end, the elderly. The contribution of aging to the immunology of populations could explain the evolutionary paradox of senescence.[41] According to the utilitarian criteria of evolution, the persistence of an aged organism that is no longer capable of reproducing has no real interest for an animal society; therefore, salmon die after reproducing and thus represent an ideal evolutionary form, the one most in accord with the model of efficient natural selection. On these extremely schematic models, then, aging does not seem to be "useful" to the group. But if forced to say why we get old, I would answer: in order to participate in the immunity of populations.

The contribution made to the immunity of populations by elderly individuals is in accord with the "grandmother theory," which emphasizes the considerable contribution made by aged women to the evolution of our species. Women are practically the only female mammals that experience menopause. With the possible exception of whales, there is no female animal living in the wild in which we see, long before the end of its life, the biological event that leads to the loss of reproductive capacities. While humans reach sexual maturity later than do the great apes, they also live longer, and therefore their reproductive capacities are in theory superior. These capacities are explained in part by the grandmother theory,[42] an evolutionary explanation of menopause proposed by Kristen Hawkes, of the University of Utah, in 1998. About two million years ago, according to Hawkes, the first hominids became very fond of high-calorie vegetable roots. Older women are supposed to have helped the younger ones with the difficult job of harvesting these roots for the children, thereby ensuring that the latter got enough to eat. In addition, by taking care of the children, the grandmothers made it easier for their daughters to carry out the task of reproduction. Another University of Utah researcher, Alan Rogers, has made a comparative demographic study of the Taiwanese population at the beginning of the twentieth century (when it was still relatively untouched by Western civilization) and the population of chimpanzees in Tanzania's Gombe National Park.[43] His conclusions support Hawkes' hypothesis: natural selection had, in the course of evolution, limited the period of the human mother's fertility in order to favor that of her daughter, whereas mother chimpanzees are useless, or rather less useful, for their daughters' maternity. The grandmother's positive role is regularly observed in Africa, where the risk of maternal mortality is as high as 1 percent. A recent demographic study of a German town in the eighteenth and nineteenth centuries reached results that are clearly in accord with the grandmother theory.[44] Whereas proponents of the grandmother hypothesis see in it only a demographic advantage for the species, it also involves a significant advantage in terms of the immunity of populations.

The influence human beings exercise on the planet has increased faster than the population itself. For example, between 1860 and 1991, the world's population quadrupled, whereas energy consumption increased hundredfold. Environmental factors (chemical pollution, the emergence and reawakening of infectious agents, etc.) corollary to demographic influence can affect immunity, which constitutes one of the main systems of adaptation to the environment. Projections made by the United Nations predict a world population of between 7.8 billion and 12.5 billion by the year 2050. In a context theoretically so favorable to the propagation of infectious agents, how should public health on the global scale be managed? New vaccinations? New drugs? Social actions? Respect for the environment?

The eradication of infectious diseases is now out of the question. Setting aside the exceptional success in the case of smallpox, efforts to eliminate malaria have been abandoned, and one suspects that the predicted eradication of measles and polio may be only idle boasts made by the heads of international organizations concerned about protecting the credibility of the institutions that provide their livelihood.

As Anne-Marie Moulin has written, ". . . the mortality rates for infectious diseases (20 percent lower than the overall mortality rates in France) are less important here than the symbolic reversal of a situation that provided grounds for optimism: in 1978, mortality rates for infectious diseases stood at about 8 percent, whereas now they are more than 10 percent, including 1 percent due to AIDS alone."[45] Since it proved impossible to drive pathogens out of the living world, it was resolved to intervene only on a case-by-case basis. For this kind of approach to be effective, we have to respond as soon as possible when there is an outbreak of infectious disease; that is, we have to monitor, and never relax our necessary vigilance. "For the dream of eradication we have therefore substituted that of a global system of monitoring, which raises its own problems," Moulin writes. "Based on greater coordination among governments and a harmonization of objectives, a program of monitoring diseases on a worldwide scale raises the delicate question of the infringement on the rights of nations." Let us also keep in mind that emergent or re-emergent pathogens do not concern the human species alone; the phenomenon is universal. Microbes attack animals as well as humans, as we currently see in the case of avian flu. Zoonoses (diseases shared by humans and other vertebrates) have completely changed the classification of human infectious diseases, and they have also changed the givens of medicine itself. There is no longer any question of eradicating microbes, but only of being vigilant. Monitoring that allows us to discern as soon as possible the arrival of a pathogen must be combined with effective biotechnological techniques for identifying it. The identification of the threat (virus? bacterium? prion?) is meaningless unless one has weapons for fighting it (vaccines, specific medications). In order to battle an epidemic, we must first discover it; second, identify the pathogen involved; and third, stop it.

Monitoring is one of the ways to avoid epidemic chaos. As was reported in the proceedings of the Fifty-fourth World Health Assembly, in 2001, "The threat to public health constituted by infectious diseases is continually evolving because of pathogens, because transmission is made easier by a changing physical and social environment, and because of the appearance of resistance to antimicrobial agents." At the global level, there is a system for fighting epidemics. Schematically, it is the WHO that coordinates international strategy, gathers information, and helps the countries affected. The information collected by the WHO comes from both official sources (laboratories and epidemiological reports) and unofficial sources such as nongovernmental organizations and various news media. There is a global network of laboratories and epidemiologists that monitors outbreaks and warns about imminent dangers, especially flu, hemorrhagic fevers, resistance to antimicrobial agents, and so forth.[46] Anyone can keep informed about infectious outbreaks by consulting the WHO's web site[47] or its "Weekly Epidemiological Record."[48]

In France, the Institut de Veille Sanitaire (InVS, "Institute for Public Health Monitoring") was created in 1998 and placed under the aegis of the Minister of Health. This institute now performs the functions of what used to be the National Network of Public Health created in 1992. Among its missions is "to detect any threat to public health and alert public authorities." Naturally, its area of competence includes infectious diseases. Its director reports that "The InVS . . . now has two hundred employees, including one hundred and fifty specialists trained in epidemiology. With a budget of 2.3 million Euros, it constantly monitors public health, studies the environment, and analyzes our lifestyles." Official medical structures such as hospitals, as well as private organizations like the Groupes Régionaux d'Observation de la Grippe (Regional Groups for the Observation of Influenza) notify the InVS of any anomalies that occur.[49]

Worldwide monitoring of the emergence of communicable diseases is provided by the WHO, which has 190 member states. This organization does not have its own technical means, but it is aided by laboratories known for their competence and by collaborating centers. The WHO has six regional offices (Copenhagen, Washington, D.C., Brazzaville, Delhi, Alexandria, and Manila) as well as national offices. It includes a Division of Emerging, Re-emerging and Other Communicable Diseases, Surveillance and Response (EMC). The WHO has set up a global monitoring project intended to spot emerging and re-emerging diseases and microorganisms resistant to drugs. The effectiveness of this network relies on increasing the collaborating laboratories' ability to diagnose infections as well as on computerizing all data (the WHONET program). Some countries have established monitoring systems focused on specific pathologies that are of particular concern locally. For example, the United States' FoodNet program is designed to monitor outbreaks of infections related to foods.

The emergence of AIDS has also led to the establishment of specific systems reflecting the magnitude and the particularity of the epidemic. Serological tests have been available since 1985, and since 1986 it has been mandatory to report the disease. Monitoring is carried out on three levels: screening, tracking the numbers of HIV-positive individuals, and health care.[50] These monitoring tools are also international, and first of all European, with more than forty countries affiliated with the WHO providing standardized epidemiological data.[51] Since the appearance of a form of Creutzfeld-Jakob disease connected with BSE (mad cow disease), an epidemiological monitoring system involving Germany, France, Great Britain, Italy, and the Netherlands has been created. This network is financed by the European Community.

Because of the complexity of the origins of epidemics, various factors have to be monitored, including climatic, ecological, and social parameters. Space technology could also provide significant help in monitoring certain diseases such as malaria. A study of malaria in Gambia, using observation satellites[52] that made it possible to examine a large area of the Earth's surface, furnished interesting information. For example, very-high-resolution radiometers installed on the satellites provided valuable data on tropical vegetation, which is directly correlated with rainfall; the latter plays an essential role in the ecology of the mosquito that is malaria's vector. The same can be said about information provided by the satellites regarding temperature, an important factor in the mosquito's growth, as well as about humidity, vegetation, and so forth. Factors such as the NVDI (Normalized Difference Vegetation Index) and CCD (Cold-Cloud Duration) are among the elements that can be evaluated long-distance and that enable us to explain spatial and temporal variations in the transmission of malaria and to contribute to the monitoring of this endemic disease. NASA has also helped develop programs for monitoring malaria in rice paddies in California's central valley, the Mexican state of Chiapas, and Belize. We have seen how *El Niño* can contribute to the outbreak of cholera epidemics, and naturally satellite observation of changes in temperature caused by the current is of great interest in monitoring efforts. The ecological approach to the emergence of epidemics should make it possible to do a better job of monitoring and probably of anticipating outbreaks. But monitoring is not sufficient. It allows us to discover the epidemic at the outset, but despite the tools that are available today or will be available in the future, monitoring will never prevent microbes from taking us by surprise.[53] Consequently, we have to act as early as possible, before the microbe has spread, in order to limit the scope of the epidemic. To do so, we have to know what infectious agent is involved. Only by identifying the attacking microbe can we respond to it specifically. To identify this invader and eliminate it, the solutions provided by scientific research and biotechnology are indispensable.

The techniques used in laboratories to study dangerous new or "emerging" germs are constantly being improved. Thanks to research carried out in the domains of immunology, virology, and so forth, we have effective tools for recognizing harmful microbes. Once a harmful microbe has been identified, it is essential not only to take the usual protective steps, but also to develop new medicines and vaccines that specifically target it. Now more than ever, it is necessary to broaden scientific investigations intended to improve our natural means of defense. We need new antibiotics and new therapies so that cultural resistance can supplement and support natural resistance. These include the "vaccines of the future,"[54] in particular vaccines composed of DNA and those used in connection with treatments in accord with the protocols of active immunotherapy.[55] Today, we have moved far beyond the time (1967) when William H. Stewart, the United States Surgeon General, declared that "the time has come to close the book on infectious diseases." That book lies wide open, and it has many pages.

The task is difficult, and microorganisms are diabolical. It is now commonplace to encounter microbes resistant to conventional treatments. That is why scientists are pinning their hopes on unconventional antibiotics. Among the latter are bacteriophages[56] and "caths" (cathelicidines), antimicrobial peptides[57] or the peptides of fish mastocytes, piscidines, and, finally—and this is a juicy project—the possibility of making bacteria produce new antibiotics.

The panoply of protection will perhaps someday include biotechnological applications allowing us to create the most elaborate forms of immunological prostheses, such as gene therapy. The latter is theoretically capable of correcting any DNA anomaly, whether congenital or acquired, and thus the genes for the immune system's MHC and HLA components, for example, could be targeted for improvement of their adaptive or cognitive defects. Gene therapy could add new characters to the genome of individuals with deficient immunity—new characters determined by the manipulator in accord with ethics and the wishes of the patients themselves, thereby giving them the power to resist one or more microbial attacks. Gene therapy has an immense potential field of application, including AIDS and other infectious diseases. The treatment of many microbial infections could benefit from modifications of the human genome and/or of those of intermediate hosts such as the *Anopheles* mosquito, the vector of malaria. We must emphasize the difficulties, both theoretical and practical, that are encountered by those who are trying to use this kind of treatment without neglecting ethical rules that should be scrupulously respected. With gene therapy, culture is modifying what is most essential in nature: its genes. Thus, an anthropocultural feedback loop is formed. Culture being a product of nature (the brain and thought), it is no surprise that genes use the former to modify the latter. Culture is the way genes have found to direct their mutations. Considering humanity's capacities for producing antimicrobial weapons, its know-how in the form of vaccines, antibiotics, and immunotherapies, and considering that the inventive faculties that have issued from scientific knowledge are constantly growing, we can say that in the not too distant future man's main immune organ will be his brain.[58]

Moreover, we should emphasize here the incontestable relationship established between the cerebral system and the immune system. The links between these two functions essential for humans' survival are fundamental and organic. Both systems are able to recognize, react, and memorize. They play a role in biological and cultural adaptation. It is thus extremely pertinent to provide for higher-level research in the domain of neuroimmunology and more generally of the mutual feedback relationships between the brain and the immune system. The importance of this research has been stressed by many experts, including Fred Rosen,[59] in an article entitled "The Lasting Lessons of SARS" (*CBR Breakthrough,* July 2003: 4–5). Completing our knowledge of the mechanisms of the relationships between the central nervous system and the immune system will certainly provide us with crucial information for "managing" epidemics, and especially for preventing them. We can only encourage the research being carried out by certain teams, such as that of Charles A. Janeway, Jr.,[60] which focuses on cerebral immuno-monitoring. It also seems pertinent to encourage research that might lead to preventive response to bacteriological attacks such as hang over our heads as a result of the threat of bioterrorism. Perhaps we should also appeal to neuroethics in approaching such serious and potentially very dangerous problems.[61]

The histories of epidemics follow a fixed scenario that leads Gaia toward chaos. They begin with the arrival in the ecosystem of a virulent microbe. The latter then spreads, flourishing on poverty, hunger, travel, commerce, various social disorders, the foibles of humanity. These histories are characterized by a sort of perpetual movement involving the microbe-human couple, the eternal accomplices in the ongoing cycle of the rebirth of infectious chaos. Examining this dialogic pair[62] whose actors are both complementary and antagonistic, and scrutinizing the behavior of the descendant of Cro-Magnon during the cataclysm, we will permit ourselves this brief maxim: Man is the epidemic, and he can face the dangers he helps to create only by means of his immunity combined with his cerebral functions. The example of the AIDS pandemic shows that in order to stop a ferocious invader perfectly well-prepared to cause catastrophic effects, the groping efforts of scientists testify not only to the hopes we can and should accord to them but also to the intellectual and material difficulties they encounter in carrying out their task. One has only to read reports on current research to realize how much intelligence, will, and material means are often required to participate in the struggle.[63]

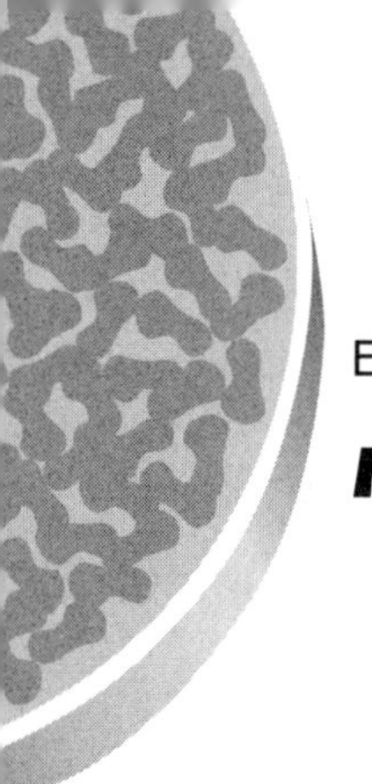

EPILOGUE

# MAN IS THE EPIDEMIC

In the ongoing, very successful drama of microbial activities, are there any new actors? Setting aside for the moment bioterrorism, which should not be confused with bacteriological warfare, the study of infections reveals certain events that are worth special attention. The preceding chapters were written several years ago, but there has been no fundamental change in the situation that would force us to alter the conclusions we have drawn. Recent outbreaks of new diseases or reappearances of old ones confirm the view that "man is the epidemic." We have seen the danger of infections arising from contaminated foods, the frequency of infections acquired in hospitals, the resistance of pathogens to the drugs that were supposed to eradicate them, and the spread of AIDS, which is now attacking Russia and China after having devastated Africa. To all these terrible consequences inseparable from human activities, we must now add the very real risk of a bioterrorism that makes use of pathogens or the toxins they produce.

## THE DANGER OF CONTAMINATED FOOD

The danger of contaminated food remains. As Claude Fischler has written, "Recently, and in particular since the first crisis of mad cow disease in 1966, the question of the safety of our food supply has become a subject of intense public attention and debate." For many years, huge pyres have burned in the English countryside as thousands of cattle suspected of carrying the fatal prion that causes a new variant of Creutzfeld-Jakob disease in humans were consumed in prophylactic flames. These cattle became "mad" after eating carcasses of sheep infected with scrapie (an ovine disease caused by prions), and then, having become suspect, they were burned to a crisp. Similar scenes were repeated in other European countries. Herds were sacrificed, both the animals and the people who raised them being victims of a greedy and irresponsible industry.

This bovine massacre did not suffice to guarantee the safety of the meat supply, however. That is why laboratory tests were set up in order to detect prions in meat intended for human consumption—but, alas, illegal sales of prohibited meat have been reported. Detection of prions in France is effective and should protect consumers, though it is hardly surprising that consumers are not entirely convinced by the reassuring speeches given by officials.

In addition to the prion, the "little newcomer" on the scene of contaminated foods, the familiar threats are still there: the bacterium *E. coli* (especially the serotype O157:H7), which is a danger in meat, and salmonella and listeria, which are usually found in eggs and cheese, respectively.[1]

Taken as a whole, precautionary measures ought to be able to prevent, so far as possible, tragedies like mad cow disease. Once again, the danger of infection by way of food supplies contaminated through human activities is not a new one; to see this, one has only to read Upton Sinclair's novel *The Jungle,* which was published in 1906 and based on fact.[2]

In concluding these remarks on the danger of contaminated food, let us once again yield the floor to Claude Fischler: "The successive crises have had as their most notable effect a radical questioning of the processes of agro-alimentary production that have been gradually established since the 1950s, and that are now criticized as characterized by 'productivism.'" These crises have led to tensions that "run throughout the current relationship between consumers and what they eat, which is increasingly transmogrified, increasingly passed through industrial 'chains,' increasingly foreign to local domestic ecosystems, produced elsewhere, who knows where or how, and distributed as an object of mass consumption." It is this notion of mass consumption—the consumers living in different countries, and often on different continents—that constitutes the danger of the contamination of foodstuffs by microbes.

### AIDS

AIDS is spreading. In 2002, the WHO estimated that 42 million people were contaminated, whether or not they had already contracted the disease. Among them were 38.6 million adults, including 19.2 million women, and 3.2 million children under the age of fifteen. In the same year, 3.1 million people with the disease died, including 610,000 children.

This brings the total number of deaths from AIDS, since the beginning of the epidemic, to almost 22 million (including 9 million women and 4.3 million children). Of the total number of HIV-positive individuals, 29.4 million live in sub-Saharan Africa, and 6 million in southern and southeastern Asia.

No one has analyzed the human causes of the spread of AIDS in Africa better than the Togolese sociologist and philosopher Sami Tchak: "To understand the AIDS problem in Africa, we have to take into account the widespread practice [of] having multiple sexual partners, which implies, at the level of a community, a major accumulation of risks. . . . With the system of polygamy, mistresses, wives, and especially clandestine forms of prostitution, men and women of all social classes are at great risk of contracting the AIDS virus. In some neighborhoods and villages, the arrival of a few prostitutes contaminated with HIV suffices to provoke within a short time a real tragedy. . . . The practice of having multiple sexual partners, the failure to use condoms, a general context of poverty, and precarious living conditions all promote the propagation of AIDS."

Responsible Africans oppose dangerous practices such as "dry sex," which causes vaginal lesions that make it easier for women to be contaminated and hence help spread the disease. Another very harmful factor in the spread of HIV in Africa is the "myth of the virgin," according to which having sexual relations with a virgin will cure AIDS. This myth has been strongly condemned by Ben Skosana, the South African minister of health.

In Asia, as in Africa, AIDS is spreading, even in countries considered authoritarian, such as Myanmar (Burma) and China. In China, prostitution and drugs are chiefly responsible for the spread of the disease.

Finally, let us review the debate concerning the origin of the human immunodeficiency virus. The first suspect was a virus affecting the green monkey, and the second, a chimpanzee carrying a simian immunodeficiency virus (SIVcpz) that was supposed to be the origin of HIV-1, which belonged to the M group. Another hypothesis that caused a scandal in the groves of academe was proposed by Edward Hooper. Hooper is a journalist for the BBC who published a long book entitled *The River,* in which he argued that the AIDS virus was (unintentionally) produced on a massive scale and administered to Africans in the form of a vaccine against polio. Hooper was not the first to suggest this; Tom Curtis, a Texas journalist, had already published a polemical article in *Rolling Stone* (March 19, 1992) under the title "The Origin of AIDS: A Startling New Theory Attempts to Answer the Question 'Was it an Act of God or an Act of Man?'" Curtis claimed that a polio vaccine produced by Dr. Hilary Koprowski of the Wistar Institute in Philadelphia had transmitted AIDS to three hundred thousand Africans in Congo, Rwanda, and Burundi between 1956 and 1960. The virus is supposed to have originated in a culture of the vaccine made using cells taken from a chimpanzee's kidney, just as the simian virus SV40 had already contaminated vaccines. Hooper's book, which is much better documented, recounts the activities of Koprowski's team, which set up, with the help of Belgian physicians, an animal house for chimpanzees in Stanleyville. The animals were killed to provide "fresh" kidney cells that enabled the team to produce a batch of vaccine called CHAT. According to Hooper, this vaccine was contaminated with the immunodeficiency virus. Hooper and Curtis's theory is contested by those whom it accuses, and studies of the AIDS virus's genome contradict the claim made in *The River*. Hooper's work shows that in 1956 researchers were aware of the simian virus, but in the race to produce a saleable polio vaccine as rapidly as possible, all caution was thrown to the winds.

## DENIAL

Recent years have also confirmed an attitude as old as the first epidemics: denial. The strongman of denial, so to speak, is Thabo Mbeki, the president of South Africa. Speaking as head of state, Mbeki has denied that HIV plays a role in the onset of the disease AIDS, thus exasperating everyone directly or indirectly associated with international public health organizations. That is why the scientists who met in Durban in 2001 for the International AIDS Conference signed the "UNAIDS Compendium on Discrimination, Stigmatization, and Denial."[3] Another UNAIDS document notes, "Denial causes individuals to refuse to acknowledge that they are threatened by a previously unknown virus which requires them to talk about, and to change, intimate behaviour, possibly for the rest of their lives. Denial also causes communities and nations to refuse to acknowledge the HIV threat, and the fact that its causes and consequences will require them to deal with many difficult and controversial subjects, e.g. the nature of cultural norms governing male and female sexuality, the social and economic status of women, sex work, families separated by migration/work, inequities in health care and education, injecting drug use." Here, UNAIDS is applying to a single epidemic what Jean Delumeau said about epidemics in general: "When the danger of contagion appears, the first impulse is to try not to see it. Chronicles dealing with plague epidemics show how leaders frequently fail to take the steps that the imminence of the danger requires."

### BACTERIOLOGICAL WARFARE

If reading the preceding lines inclines us to criticize political leaders, what must we think of government leaders who have envisaged bacteriological warfare? The latter situation is not new. The spread of the Black Death across Europe resulted from the use of bacteriological warfare during the conflict between the Genoese and the khan of the Mongolian Golden Horde. In 1345, following a disagreement between the Mongols and the Genoese, the former laid siege to Kaffa (modern Feodosia), a city in the Crimea that was then a Genoese trading post. Despite the length of the siege (three years), the khan, Janis Beg, was certain he would eventually succeed until he saw his troops begin to melt away, decimated by plague. Irked, before lifting the siege he used catapults to throw the bodies of plague victims inside the city walls. The Genoese, freed but contaminated, left the Crimea, taking the devastating microbe along with them when they returned to western Europe. The khan's tactic was adopted by other leaders: in 1710, the Russians projected the bodies of plague victims onto the Swedish army, and in 1785, in La Calle (Tunisia), the Nadis tribe catapulted the clothes of plague victims over the city walls so that they would fall on the besieged Christians. In Canada, during the French and Indian War, the British commander, Lord Jeffrey Amherst, ordered Colonel Henry Bouquet, who was under Indian attack at Fort Pitt (now Pittsburgh) to withdraw, deliberately leaving behind blankets contaminated with the smallpox virus so that they might be found by the Indians, who were allied with the French. In memory of this tragic episode, the Quebec artist Daniel Castonguay has created a moving work representing three Indians wearing red blankets contaminated by the smallpox virus. The future victims, their heads bowed, mourn a member of their tribe who has died of smallpox. These blankets were Trojan horses carrying a militarized virus.

Much more recently, during the Cold War, the Russians and the Americans accused each other of resorting to bacteriological warfare in Korea and Afghanistan. No proof was provided in either case.

People have always feared bacteriological warfare. In 1931, the Japanese built factories in Manchuria to make biological weapons. Per month, these factories produced up to three hundred kilograms of plague bacillus, a ton of various pathogenic bacteria (cholera, typhus, dysentery), and forty-five kilograms of fleas. The best-known of these factories, Unit 731, used human guinea pigs, thousands of whom are said to have died. Drawing on the macabre output of this factory, the Japanese used aircraft to drop the plague bacillus on Chinese villages, causing seven hundred deaths in Nimpo, to give just one example. The Japanese authorities responsible for these atrocious war crimes provided their American counterparts with intelligence in exchange for their silence and complicity. Nonetheless, this kind of thing led to the signing of the Biological and Toxic Weapons Convention on April 10, 1972, with Great Britain, the USSR, and the United States as trustees. Signatory states pledged not to develop, produce, stockpile, retain, or acquire bacteriological or biological weapons or toxins. In reality, however, the Russians and the Americans were duping each other, for both of them were—like many other countries, moreover—secretly pursuing the production of biological weapons in special research laboratories. The American research program and the development of biological weapons were described in a book by Judith Miller, *Germs: Biological Weapons and America's Secret War* (New York: Simon & Schuster, 2001). So far as the Russians are concerned, two serious accidents revealed their involvement in bacteriological warfare. In 1971, on the island of Vozrozhdeniya (in the Aral Sea) a military accident led to contamination with the smallpox virus. As we mentioned in an earlier chapter, another accident that occurred in 1979 in Sverdlovsk proved that the Soviets were engaging in doubletalk. At the time, Soviet authorities blamed the sixty-eight civilian deaths on a shipment of rotten meat. Today, we know that there was an epidemic of anthrax,[4] the bacillus having entered the victims' bodies through the lungs after escaping from a military complex where seven thousand people were employed.

In 1999, Kanatjan Alibekov, a Russian bacteriologist who had defected to the United States, published under his new name, Ken Alibek, a book entitled *Biohazard*. According to this presumably well-informed author, the Russian laboratories in which biological weapons were produced and which were known by the generic name of the project, "Biopreparat," were still operating in 1998. He says that in 1988—that is, smack in the middle of the Glasnost era—Soviet military leaders were planning to put four hundred kilograms of anthrax spores into SS-18 missile warheads. The anthrax bacillus is a choice agent for bacteriological warfare, because it forms spores that are able to survive in various environments. However, the spores tend to aggregate rather than remaining in suspension in the air. Perverse scientists have been able to "militarize" these spores by disaggregating them, thus making them more easily disseminated through the air and more likely to cause the pulmonary form of anthrax, which is, as we have noted, far more lethal.

In addition to anthrax, among the microbes preferred by strategists of bacteriological warfare are those associated with smallpox, plague, Q fever, tularemia, viral encephalitis, hemorrhagic fevers, and botulism. Genetic engineering has now produced genetically modified microbes into which genes from other microorganisms have been introduced in order to make the transformed microbes even more dangerous. The 1972 Biological and Toxic Weapons Convention was signed by 114 countries.[5] Although some of the thirty-two countries who are not signatories pose no threat, such as Andorra, others are potentially much less innocuous. According to a CIA study issued in 1995, seventeen countries seemed to be disregarding the convention and constructing—or at least trying to construct—bacteriological arsenals.[6] If we believe Hussein Kamal, an Iraqi general who defected in 1995, his country had produced "19,000 liters of concentrated botulin toxin, 8,500 liters of anthrax spores, 220 liters of aflatoxin, as well as the gangrene bacterium, the yellow fever virus, dromedary variola, etc." The Biological and Toxic Weapons Convention is, thus, far from being able to protect us against all bacteriological warfare.

It is hard to know just how seriously to take the danger of bacteriological warfare at the present time. In any event, it is clear that current concern is focused above all on bioterrorism.

## BIOTERRORISM

After the September 11, 2001, attacks in New York and Washington, a letter contaminated by anthrax made front page headlines in the United States and once again shook the American people's belief that they were safe. How can one fail to be astonished by the deafening silence of the American media, which for some strange reason failed completely to pursue the results of the investigation into the search for the perpetrators? Conspiracy theorists will suggest that the person or persons to blame are known but that naming names would be very embarrassing for American leaders. The first anthrax attack occurred in Florida on October 4, 2001: it was learned that Bob Stevens, a journalist, had a respiratory infection caused by the bacillus. The possibility that this had a "natural" cause was quickly set aside in order to assign the blame to members of Al Qaeda who were supposed to have sent contaminated letters.[7] Subsequent events (letters intercepted in New York and New Jersey, potentially affecting postal employees) confused matters further, though it was clear that those responsible wanted news coverage, because the letters containing the bacteria were sent to the media. Then it was reported that the anthrax in question came from a laboratory in Ames, Iowa, that was working on "biodefense projects." People gradually accepted the notion that the anthrax sent through the mails was a form of domestic terrorism. As of now, we still don't know who the diabolical senders were.

Curiously, the American public has also forgotten that in 1984 the American Rajneesh sect, based in Oregon, had already contaminated a salad bar in The Dalles, near Portland, with salmonella. Let us mention the more recent Japanese Aum sect's 1995 attack on the Tokyo subway using the lethal sarin nerve gas. This was a chemical attack carried out by fanatics who had earlier unsuccessfully sought to "militarize" the botulin toxin and anthrax bacteria. They encountered the main difficulty faced by terrorists: making microorganisms dangerous in the environment where they are to be disseminated. It is in fact difficult to condition infectious agents in such a way as to be able easily to use them on the site of a terrorist act.

In other words, should we really fear biological terrorism? According to experts, although the threat should not be underestimated, it remains limited because of the problems connected with the use of biological weapons. The four absolute imperatives for the use of such weapons are: producing the terror agent, conditioning it so that it can be transported without losing its virulence, providing the means of getting it to the site (plane, missile, artillery shell), and the means for dispersing it (explosion, spraying). For terrorists, biological weapons have a few advantages. Producing them does not require excessively complex or expensive equipment, the harmful effects on populations are potentially considerable, they can easily be concealed (invisible, odorless, tasteless), and their cost effectiveness makes them excellent "poor people's weapons." In practice, it would therefore be wrong to neglect the threat, because terrorists can gain access to money, scientists, and complicitous governments. Since the anthrax cases that occurred in Boca Raton, Florida, shortly after the September 11, 2001, terrorist attacks on the World Trade Center and the Pentagon, the biological threat has been feared in the United States, and this fear was exacerbated by the diffusion of contaminated mail. In 1998, a campaign to vaccinate American soldiers against anthrax was launched, and the damage done by infected mail accelerated improvements in the preparedness, particularly rapid diagnostic tests to determine the nature of the agent used in an attack. In addition to the conventional means of protection (gas masks, shelters, special clothing, decontamination, vaccination, specific drugs), it is essential to detect and identify the microbe used by terrorists.

Among the most feared infectious agents that could be used by bioterrorists (smallpox, anthrax, plague, Ebola, botulism, tularemia), smallpox comes out far ahead.[8] That is why President George Bush announced on December 12, 2002, the decisions made by his administration regarding smallpox vaccination. The objective was to vaccinate five hundred thousand soldiers during the summer of 2003. The long-term goal is to vaccinate ten million people (healthcare personnel, firefighters, police). However, mass vaccination against smallpox has been discontinued in the United States, and neither the president nor his family will be revaccinated. These decisions were made after many hesitations. Theoretical studies of models of smallpox transmission have stressed the value of mass vaccination, but the side effects of immunization are not negligible.[9] To better anticipate the effects of an attack, a simulation called "Dark Winter"—representing the United States during a bioterrorist attack with smallpox—has been elaborated. President George Bush also ordered a videocassette copy of a BBC docudrama (*Smallpox 2002: Silent Weapon*) in which a "suicide patient" wandering around New York triggers a worldwide pandemic that results in sixty million deaths. There is no doubt that the fear of bioterrorism contributed to the creation of the Department of Homeland Security.

An epidemic of hoof-and-mouth disease in 2001 showed how greatly a bioterrorist attack on livestock and/or agricultural produce could affect a country's economy. In 2001, four million cattle were killed in Great Britain (still more sacrifices), and the economic repercussions of mad cow disease have been put at nearly fifty billion dollars. The mad cow epidemic was not a bioterrorist attack, but the catastrophe provides a good illustration of what that kind of attack might achieve. However, agricultural bacteriological warfare is not new, either. During World War I, the Germans are supposed to have tried to contaminate the Allied armies' horses with "glanders," a contagious bacterial (*Pseudomonas*) disease fatal to horses, and in 1916, in southwestern France, German saboteurs were accused of having contaminated with the anthrax bacillus cattle imported from Argentina. The danger of agricultural bioterrorism seems all the greater because the species threatened (animals raised for profit and plants) are genetically selected and therefore all have the same genetic background. Hence if we assume a certain susceptibility, all the individuals involved will be at risk in a textbook example of a deficient immunity of populations. The danger, particularly in developed countries, is not so much famine as economic problems. Moreover, one cannot exclude the possibility that infectious diseases could be transmitted from animals to humans. In short, the risk seems so real that in its 2003 budget, the United States government allocated a supplementary sum of 146 million dollars to anticipate and prevent agricultural bioterrorism. Agriculture itself can provide biological weapons for terrorists, such as ricin, a by-product of castor oil that is intended for industrial use. Ricin is a potent poison that can be added to food or drinking water.

## CONCLUSION

In June 2001, the American journal *Science* published a humorous editorial written by two eminent biologists. This was a commentary on the general meeting of a fictitious society called the "World Pathogen Association," presided over by a prion. In the course of the meeting, the president informs his microbial colleagues that man, the dominant species on Earth, has replaced their "natural" prey, so that pathogens are confronted by a challenge: they have to adapt to their new terrain. "Homophagia is the way of the future" becomes their slogan. The prion-president goes on to laud pathogens that have been able to adapt to their victims. Among the most successful, he says, are the HIV virus, tuberculosis bacilli, the Ebola, Hanta, Lassa, and Marburg viruses, the pathogens that cause Legionnaires' disease and malaria, the dengue fever virus, the spirochete that causes syphilis, and (to end this gloomy list), the flu virus. His closing remarks remind the microbes that they have grounds for hope (and it's their president who is speaking), and the main one is man himself and his behavior. We share—with regret, to be sure, but without hesitation—the microbes' opinion: "Man is the epidemic."

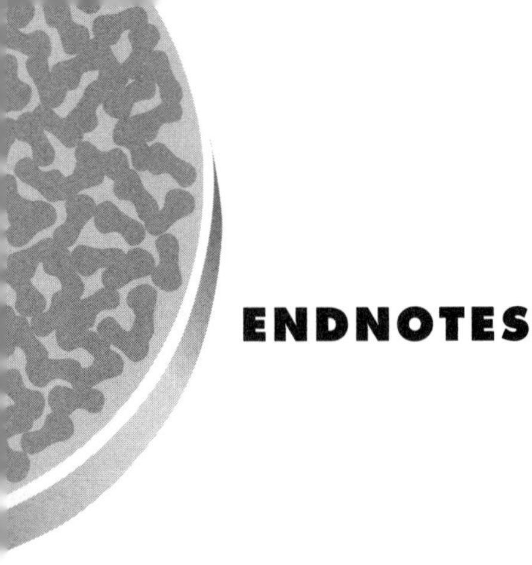

# ENDNOTES

## NOTES TO CHAPTER 1

1. "Nana remained alone, her face turned up, in the light of the candle. She was a charnel house, a pile of humors and blood, a shovelful of corrupted flesh, thrown there on a cushion. Pustules had taken over her whole face, one touching the next; shriveled, shrunken, gray like mud, they already seemed to be kind of earthy mold on this shapeless mess on which one could no longer distinguish the features. One eye, the left one, had completely sunken into the boiling purulence; the other, half-open, plunged into a dark, rotten hole. The nose was still oozing. A massive reddish crust started on one cheek and overran the mouth, which it distorted into a horrible smile. And on this horrible, grotesque mask of nothingness, the hair, the beautiful hair, retained its sunny brilliance, fell down in a shower of gold. It seemed that the virus she had caught in streams, from the carrion allowed to lie there, this ferment with which she had poisoned a people, had just risen into her face and rotted it."
2. The smallpox virus is one of the most complex infecting our species, and only human beings become ill from it.
3. The immunology of populations is an approach to the relationships between human groups and the microbial environment, interactions between the means of defense represented by the immune (super)system, species, and microorganisms or any other elements that can affect a population's immune response. The immunology of populations is not solely the result of the emergence of biological phenomena, but is also produced by cultural adaptation.
4. B. Jahrling, a virologist working in a U.S. Army laboratory at Fort Detrick (Maryland) tells of having received from a Russian colleague (Lev S. Sandakhchiev, director of the Vektor research institute near Novosibirsk) a video cassette showing, in a Siberian village, "the exploration" of human bodies buried in

the tundra centuries ago. These were the bodies of people who had died of smallpox in the Siberian province of Gorno-Altayskya. Considering the stability of the virus, we can see that these bodies constitute an inexpensive "lode" of material for making a potential biological weapon.

5. In May 1996 delegates to the WHO from 190 countries adopted a resolution stipulating that "the remaining stocks of variola virus, including all whitepox viruses, viral genomic DNA, clinical specimens and other material containing infectious variola virus, should be destroyed on June 30, 1999 . . ."
6. In November 1997, an epidemic of monkeypox broke out in Kasai province of the Democratic Republic of Congo (the former Zaire), resulting in thirty deaths. This event followed outbreaks of the disease in the same country, in July, August, and September 1997, after an appearance of this infection at the beginning of the year 1997. The worrisome aspect of these epidemics is that the virus is transmitted to victims not only by monkeys (and squirrels) but also by other human victims. In other words, the high rate of inter-human transmission indicates that the virus is well adapted to man, which increases the risk of future epidemics.
7. In the *Decameron*, Boccacio tells how, during a plague in Florence, his heroes fled (and it was wise to do so) to Fiesole, on the heights above the city. The story of the epidemic is a faithful description of the scourge (in Latin, *pestis* means scourge): "The plague described here was of such virulence in spreading from one person to another that not only did it pass from one man to the next, but, what's more, it was often transmitted from the garments of a sick or dead man to animals that not only became contaminated with the disease, but also died within a brief period of time." Boccacio, The *Decameron*, trans. M. Musa and P. Bondanella (New York: Norton, 1977), 5.
8. "Its ravages went far beyond anything produced by ordinary diseases . . . In France, the first wave (1348–1349), which crossed the whole country from south to north, was disastrous: depending on the location, a fourth, a third, half, and sometimes as many as 80 or 90 percent of the population died."
9. "So God came to David and told him, and said to him, 'Shall three years of famine come to you in your land? Or will you flee three months before your foes while they pursue you? Or shall there be three days' pestilence in your land?'" David chose the plague, "So the Lord sent a pestilence upon Israel from the morning until the appointed time; and there died of the people of Beer-Sheba seventy thousand men." 2 Samuel 24: 13–16.
10. "But the city itself began now to be visited too, I mean within the walls; but the number of people there were indeed extremely lessened by so great a multitude having been gone into the country; and even all this month of July they continued to flee though not in such multitudes as formerly in August, indeed, they fled in a such a manner that I began to think there would be really none but magistrates and servants left in the city." In Defoe, D. *A Journal of the Plague Year*. London: Penguin Books, 1986.

11. Genoese galleys arrived in Marseilles on All Saints' Day 1347. They came from Crimea, and had spread the disease in Constantinople and Messina to the point that Livorno (Leghorn) and Genoa had denied them entry to their ports. The authorities in Marseilles, thinking they knew better, allowed the ships to land, with the well-known consequences.
12. During the plague of 1585, Montaigne, who was then mayor of Bordeaux, took refuge on his country estate. In July, two officials who had stayed in Bordeaux wrote to ask him to return to the city. On July 30, they received the following disappointing reply: "On Wednesday I will come as near you as I can . . . to Feuillas, if the disease has not arrived there."
13. On the occasion of each epidemic, people looked for scapegoats. Their blind and unfounded vindictiveness was usually directed toward minorities, especially if they appeared to be, in one way or another, heterodox. Jews and lepers paid a heavy price for their neighbors' ignorance. In 1348, throughout Europe Jews were burned and their property was confiscated and handed over to the cities and the churches . . . "In Mainz, twelve thousand Jews were burned in a few days, and the fire was so intense that it melted the lead of the stained-glass windows in the church of Saint Quirus, near which the stakes had been set up."
14. When confronted by the danger of infection, the general reaction was to flee. The sick and the dead were abandoned. Many of the dead were found at home, in their beds or on the floor. The decomposed cadavers were shoveled up. It was fortunate when the deceased were not simply thrown out the windows.
15. "In Marseille, all 150 Franciscan monks died, as did all those in Carcasonne and all the Augustinians in Avignon; in Maguelonne, out of 160 monks, there remain only 7, and in Montpellier, out of 140, only 7 as well. The secular clergy is somewhat less affected; in Bordeaux, out of 20 canons at Saint-Suerin, 3 survive, and in Paris, of the 136 nuns of Sainte-Catherine there were only 36 deaths as compared with 4 or 5 in the preceding years."
16. "Thus in Bordeaux a girl of 16 who had inherited the property of her uncles and aunts, of her parents, and of her four brothers and sisters, as they died one after the other, and was thus a rich heiress, married in 1349 a young lord who wanted to regild his coat of arms." In Delumeau, J. and Lequin, Y. *Les malheurs des temps. Histoire des fléaux et des calamites en France.* Paris: Larousse, 1976, 177–192.
17. *Vibrio cholerae* was discovered by the German bacteriologist Robert Koch in 1884. It has a flagellum and secretes a substance that causes diarrhea.
18. At that time, a physician to Queen Victoria, John Snow, grasped the relationship between water and the disease. He saw that a well in Broad Street was contaminated by sewer water, and had the pump's handle removed to prevent the propagation of the infection.
19. *El Niño* (in Spanish, "the little one," understood here as "the child Jesus") causes a warming of the Pacific Ocean. Also called the *El Niño Southern Oscil-*

*lation* or ENSO, this phenomenon occurs every two to seven years. It is part of the summer-winter climatic cycle and affects countries on the Pacific by causing droughts in India, Australia, Indonesia, and parts of Africa and China, while at the same time leading to flooding in the central and equatorial areas of the eastern Pacific. The cooling that usually follows *El Niño* is called *La Niña*. A great deal is known about ENSO, but that does not necessarily mean that all the perturbations it may cause are predictable. The unpredictability of climatic conditions will be discussed in Chapter 5.

20. Between 1987 and 1990, ten observation stations were set up in Bangladesh along two rivers and in eight marshes 46 kilometers south of Dhaka. Samples of water and plankton were taken every two weeks. It was noted that the numbers of *Vibrio cholerae* increased in tandem with those of the copepods that contaminated fish by parasitizing them. In the form of a constituent of the commensal flora, the cholera vibrio can be carried over months and for thousands of kilometers.
21. Deaths in Chaillot street in Paris constitute a case of difference in mortality rates that is hard to understand. During the epidemic in Paris, forty-one residents on the odd-numbered side of the street died, while only five on the even-numbered side died. There are two possible explanations for this strange fact. One is sociological: the odd-numbered side is lower class, and the houses are more densely populated and less healthy than those on the even-numbered side. The other explanation is bacteriological: the water feeding the pumps on the side with a high mortality rate took their water from the Seine, which was contaminated, while the pumps on the even-numbered side took their water from the Ourcq canal, which was not contaminated.
22. The distribution of the ABO groups among cholera patients in the Philippines, with vibrio in stools (V+) or without vibrio in stools (V−). The figures indicate the percentage of patients with the disease in each group.

| ABO Groups | Total Patients | Patients V+ | Patients V− |
|---|---|---|---|
| 0 | 45.4 | 64.3 | 63.6 |
| A | 25.8 | 10.2 | 15.2 |
| B | 22.6 | 21.2 | 21.2 |
| AB | 6.2 | 3.3 | — |

23. "Despite numerous treatment rooms that were set up everywhere to receive the ill, beds and stretchers were soon all in use and the rooms were full. Whole families turned up for consultations in order to be hospitalized. Cyanotic, dyspneic persons, fearful and sometimes dying, could not be sent away, they had to be put on mattresses on the floor, in hallways or at the doors of the rooms while they waited for a bed. A certain number of them died after several hours in the hospital. Several deaths in a family were not rare: a patient might

have just lost her husband and her mother, and still have at home several other relatives less ill. We found ourselves truly faced with a great epidemic, and felt completely powerless before it. Some, seeing the gravity of the symptoms and the rapid progress of the disease, went so far as to think of plague or cholera. However, it was indeed influenza, and we were never to find in our cases complications or facts that had not already been mentioned in earlier epidemics." In *La Grippe de 1918 à 1919 dans un service de l'hôpital Saint-Antoine,* by Dr. Marguerite Barbier, quoted by J. J. Becker.

24. Epidemics or pandemics of influenza since the nineteenth century have been observed every ten to fifteen years, as is shown by the following table.

| Year | Aspect | Sub-type |
|---|---|---|
| 1889–1890 | pandemic | |
| 1900–1903 | epidemic | |
| 1918–1919 | pandemic (Spanish flu) | $H_1N_1$ |

25. The type A virus responsible for the Spanish flu epidemic of 1918 was identified as the sub-type $H_1N_1$. In the interim, this particular virus has not been found in humans. It can also infect swine, but not birds [453]. Since the 1918 epidemic, the $H_1N_1$ virus of the time has undergone a genetic drift; that is, some of its genes have gradually been modified. When the Asian type A virus ($H_1N_2$) emerged in 1957, A ($H_1N_1$) disappeared, as did $H_1N_2$ when the Hong Kong A virus ($H_3N_2$) appeared in 1968. Since 1977 and the reappearance of another $H_1N_1$, we have seen cohabitation of the latter with $H_3N_2$.
26. After Reed died during an appendectomy in 1901, Walter Reed Hospital in Washington, D.C., was named in his honor; Finlay was appointed Cuba's chief sanitation officer, and after his death the Finlay Institute for Investigations in Tropical Medicine was created in his honor by the Cuban government.
27. "Let us suppose that a disease that is frequently lethal, such as plague, attacks a human group, and let us say that out of every hundred persons who contract the disease, only twenty-five survive. To what do the latter owe this privilege? They adapted . . . Acting on a human group, the disease has carried out a process of selection. It has given rise to an appropriate protective reaction in individuals who had the relevant potentialities, which are moreover fortuitous, and which those who succumbed to the disease did not have to the same degree." In Bordet, J. *Infection et immunité.* Paris: Flammarion, 1947.
28. De Vries and his collaborators studied twenty-six polymorphic genetic traits including the complementary alleles and the Gm, HLA, and glyoxylase systems.
29. Myxomatosis in Australia illustrates the conjoint evolution of the host and the microorganism, showing how rabbits with reduced susceptibility replaced the original population, and in what direction the virus evolved. The pathogen,

of Brazilian origin, was introduced into Australia in 1950, at a time when the wild rabbit (*Oryctolagus*) population was in the hundreds of millions, animals that had descended from a dozen pairs brought to the continent in 1859. In 1960, 99 percent of these animals were dead, the demographic collapse leaving only animals resistant to the disease. At the same time, the virulent form of the virus disappeared, and was replaced by a less harmful variety. In reality, the virus gradually shifted from the very virulent type I (in the 1950s) to the less aggressive type II (1958–59), after which the still less harmful types III and IV appeared and became the most widespread (1963–64). It seems that prior to 1950 the virus had evolved in a direction that led to the complete elimination of the host necessary for its continued survival.

30. According to Mornet, it is possible to determine the demographic effects of the plague epidemic in Aquitaine. In summary, the mortality rates reported are as follows: Bordeaux, ten to twelve thousand (out of forty thousand inhabitants); Périgueux, 25 to 40 percent; Bergerac, 25 to 40 percent; Bazas, 90 percent.
31. Physicians also believed in the divine origin of the plague. At the Sorbonne, it was explained this way: "We do not wish to fail to say that when the disease proceeds by divine will, we have no advice to give other than to humbly call upon that will itself."
32. All the measures taken were not senseless; in 1465 the Sicilian port of Ragusa began quarantining ships suspected of carrying disease, and Venice did the same in 1485.
33. Epidemics frequently gave rise to xenophobia. In southwest France, Jews were massacred in Périgueux, and they were forbidden to touch the bread and fruit they bought. Guards at the city gates kept foreigners from entering.
34. In Paris, Transnonain street got its name from the verb *transnoniser* (massacre) after Jews were killed there in 1348. In early 1349, the people of Strasbourg, who were unhappy with magistrates who opposed the massacres, replaced them. On February 14 of the same year, nearly one thousand Jews—men, women, children, and old—were burned in a ditch. The survivors were forced to leave the city promptly.
35. In Toulouse, a man named Cadoz was accused of spreading the plague and decapitated in 1545. In Milan, two men named Piazza and Mora were subjected in 1630 to an even more horrible fate: they were broken on the wheel before being hanged and then burned. The fate of the two men was all the more cruel because the first, Guglielmo Piazza, a health officer, was wrongly accused and questioned under torture. The unfortunate man confessed whatever they wanted, and accused his barber, Giangiamo Mora, of making the contaminated grease. Not only were the two men tortured to death, but the barber's house was demolished and replaced by a column (the column of infamy) erected in memory of this sinister episode. Most of the European "greasers" died in anonymity and under the most atrocious circumstances.

36. Films directly or indirectly related to plague include the following:

| Author | Title | Year |
|---|---|---|
| I. Bergman | *The Seventh Seal* | 1957 |
| L. Buñuel | *Leonor* | 1975 |
| E. Castellari | *Keoma* | 1976 |
| J. Clavell | *The Last Valley* | 1971 |
| M. Cloche | *Monsieur Vincent* | 1947 |
| R. Corman | *The Mask of the Red Death* | 1964 |
| J. Demy | *The Flute Player* | 1972 |
| P. Greenaway | *The Baby of Macon* | 1993 |
| W. J. Has | *The Tribulations of Balthazar Kober* | 1988 |
| F. Lang | *The Plague in Florence* | 1919 |
| F. Merle | *Le cimetière des morts vivants* | 1968 |
| F. Murnau | *Nosferatu* | 1922 |
| L. Puenzo | *The Plague* | 1992 |
| O. Preminger | *Forever Amber* | 1947 |
| K. Russell | *The Devils* | 1970 |
| P. Verhoeven | *Flesh and Blood* | 1985 |
| V. Ward | *Navigator* | 1988 |

## NOTES TO CHAPTER 2

1. In 1979, a physician caring for the gay community in Los Angeles was intrigued by certain of his patient's problems. The clinical description of AIDS was not established until 1981. The delay is explained by the fact that it takes time for the virus gradually to erode the immune system to the point that it is irreversibly destroyed. Patients with AIDS suffer repeated infections provoked by germs that are normally not pathogenic, known as opportunistic.
2. Biological samples such as serum, the parts of the body removed in operations, and the results of autopsies relating to some of these patients were preserved, which allows us to study them retrospectively, guided by the ideas and know-how acquired through investigation of "recent" cases. The presence of HIV has been demonstrated in biological samples taken from a patient hospitalized in 1959.
3. The AIDS virus is African in origin, and its diffusion throughout the world is recent, at least after 1954. What caused the breakdown of a balance? The explanation is all the more difficult because there are two types of HIV, $V_1H_1$ and $V_1H_2$, both of them closely related to lentiviruses, which are retroviruses. $V_1H_1$ is the virus found in Central Africa, North America, and Europe. $V_1H_2$, on the other hand, is found in India and in West Africa.
4. In 1950 construction of the Belem-Brazilia superhighway began, its route traversing the Amazon forest. In the blood of people working on this project,

virologists from the Rockefeller Institute discovered unusual, little-known or even unknown viruses. In 1961 an epidemic struck eleven thousand persons in Belem. The vector of the disease was a fly that proliferated in the piles of pods from cacao trees planted by the new colonists who were following the route of the new highway. The fly had transmitted the Oropouche virus from animals to humans, from the forest to the city.

5. Wolfe, N. D. "Naturally acquired simian retrovirus infections in Central African Hunters." *The Lancet* 363 (2004): 932–936.
6. Gao, F. et al. "Origin of HIV-1 in the chimpanzee *Pan troglodytes troglodytes.*" *Nature* 397 (1999): 436–441.
7. Fauci, S. "The AIDS epidemic. Considerations for the 21st century." *The New England Journal of Medicine* 341 (1999): 1046–1050.
8. Cohen, J. "AIDS virus traced to chimp subspecies." *Science* 283 (1999): 772–773.
9. Wain-Hobson, S. Joint session of the Académie des Sciences and the Académie Nationale de Médecine. Public session, Tuesday, June 1, 2004.
10. The colonization of the former Belgian Congo, which later became Zaire and then the Democratic Republic of Congo, illustrates a situation in which the breakup of a social organization, customs, and cultural practices produced consequences from which the country still suffers deeply. In the 1880s, the colonial administration set up in the Congo, whose capital was then called Leopoldville, a system of forced labor under which whole villages were wiped off the map and their inhabitants dispersed, men and women often being sent separately to different parts of the country, and especially to work on rubber plantations. As a result, the population of Congo is reported to have fallen from thirty million in 1890 to eight million in 1924. The separation of men from women resulted from the wish to put workers into camps (which cost less than moving whole families). The De Beers company and the Mining Union of Upper Katanga had a pressing need for labor to mine gold, copper, and diamonds. This led to a breakup of the family group and thereby to a breakup of the social structure and had the effect of encouraging prostitution—all ills from which the country has suffered ever since.
11. AIDS was, and remains, a disease spread by syringes under certain conditions. This is the case for drug users who have a habit of sharing the same needle as part of a commonly practiced ceremony. In a similar way, the disease was spread in African hospitals in which extremely limited budgets sometimes led to the use of the same needle for several patients.
12. Physicians and epidemiologists at the WHO and other organizations that have worked in Africa are often discouraged by three kinds of behavior which are, alas, too common there. First, wars, pillaging, massacres, rapes, and vandalizing are virtually permanent in certain countries on the continent. The recent events in Rwanda, Burundi, and the Democratic Republic of Congo undoubtedly attest to this fact. In the Tanzanian city of Kasensero, half

the population died and almost all the survivors were contaminated after repeated attacks by Ugandan "soldiers," most of whom were HIV-positive, and behaved like bands of pillagers left to their own devices. To this factor that ruins and depopulates whole regions whose inhabitants flee, we must add the stubbornness of leaders who deny the reality of the epidemic in their countries. One example among many others: in 1988 Zimbabwe's Ministry of Health was put in the hands of a general who immediately falsified the official figures relating to AIDS deaths and forbade mentioning the disease on death certificates. Finally, last but not least, we have to acknowledge that there is a form of incredulity and fatalism among human groups: in Uganda, AIDS was ironically transformed into Acha Iniue Dogdego Scichi ("Let it kill me because I will never give up young women"), while in Zaire it was transformed into "Imaginary Syndrome to Discourage Lovers."

13. WHO statistics for December 1997 (in millions) are as follows:

| | Persons with AIDS in 1997 | Bearers of the HIV virus (sick or not) in 1997 | Deaths (1997) |
|---|---|---|---|
| Adults: | 5.2 | 29.5 | 1.89 |
| Women: | 2.1 | 12.1 | 0.824 |
| Younger than 15: | 0.59 | 1.1 | 0.46 |
| Total: | 5.8 | 30.6 | 2.35 |

14. Tri-therapy treatment costs a minimum of ten thousand to thirty thousand dollars per year. Developing countries can on average devote no more than one to ten dollars per person per year to health care. Hence Thailand seems privileged in being able to spend thirty dollars per year per person.
15. Kaldy, P. "Sida: des médicaments novateurs." *La Recherche* 364 (2003): 56–57.
16. Cohen, S. "AIDS in Ouganda: the human-rights dimension." *The Lancet* 365 (2005):2075–2076.
17. Cohen, J. "Asia and Africa: on different trajectories?" *Science* 304 (2004): 1932–1938.
18. Perry, A. "When silence kills." *Time,* June 6, 2005: 56–59.
19. AZT is known as zidovudine, 3TC as lamivudine, D4T as stavudine, and NVP as nevirapine. AZT, 3TC, and D4T are nucleosidics that inhibit reverse transcriptase, an enzyme that is necessary for the virus's reproduction. NVP is also an inhibitor, but not a nucleosidic. Often these inhibitors are combined with others that are known as "carriers" and that disrupt the final phase of the virus's replication.
20. The study carried out in Nairobi examined 498 prostitutes who had unprotected intercourse. Their serological development was compared with that of 232 Kenyan women who were not prostitutes.

21. MacDonalds K. S. "Influence of HLA supertypes on susceptibility and resistance to immunodeficiency virus type 1 infection in homosexual men." *Journal of Infectious Diseases* 181 (2000): 1581–1589.
22. This work studied, from 1993 to 1996, 171 children born to HIV-positive mothers. MacDonald, K. S. "The HLA-A2/6802 supertype is associated with reduced risk of perinatal human immunodeficiency virus type 1 transmission." *Journal of Infectious Diseases* 183 (2001): 503–506.
23. Balter, M. "Revealing HIV's T cell pass key." *Science* 280 (1998): 1833–1834. Kwong, P. D. "Structure of an HIV gp120 envelope glycoprotein in complex with the CD4 receptor and a neutralizing human antibody." *Nature* 398 (1998): 648–659. Wyatt, R. "The antigenic structure of the HIV gp120 envelope glycoprotein." *Nature* 393 (1998): 705–711.
24. Libert, F. "The CCR5 mutation conferring protection against HIV-1 in Caucasian populations has a single and recent origin in Northeastern Europe 399–406." *Human Molecular Genetics* 7 (1998): 399–406.
25. Duncan, C. J. "What caused the black death." *Postgraduate Medicine Journal* 955 (2005): 315–320.
26. O'Brien, S. "Pourquoi certaines personnes résistent au sida?" *Pour la Science* 240 (1997): 82–89.
27. Galvani, A. P. "Evaluating plague and smallpox as historical selective pressure for the CCR5-32 HIV-resistance allele." *Proceedings of the National Academy of Sciences of the USA* 100 (2003): 15276–15279. Galvani notes that the CCR5-32 allele is absent among Africans, Asians, and Amerindians, and that the mutation is recent and its frequency in Europe is high, leading some researchers to suggest that its persistence is due to the selective pressure exercised by plague. But by 1750, plague had disappeared from Europe, whereas smallpox continued its periodic ravages, striking above all children who had not yet acquired their accidental antismallpox immunity. Mortality remained about 30 percent. Galvani's age-structured model indicates that "even the heavy mortality during the Black Death and Great Plague pandemics, combined with a series of intermittent epidemics, does not generate sufficient selective pressure to drive a resistance allele to 10 percent frequency. Instead, a disease with relatively high case fatality rates that persisted more continuously since the origin of the allele was likely to have been responsible. Smallpox is such a disease. No single smallpox pandemic was as devastating as the Black Death, but the cumulative toll of human life caused by smallpox constituted an even stronger selection pressure than the episodic decimation of bubonic plague." Other researchers think it probable that the selection of the CCR5-32 trait by the smallpox virus was determinant because of the frequency of the virus's mutation. Cf. A. P. Galvani, "The evolutionary history of the CCR5-32 HIV-resistance mutation." *Microbes and Infection* 7 (2005): 301–308.
28. Brewer, D.D. "Mounting anomalies in the epidemiology of HIV in Africa: cry

the beloved paradigm." *International Journal of STD & AIDS* 14 (2003): 144–147.

29. Important differences have been observed between the form of HIV infecting different patients and possibly even the same patient, to the point that "no HIV virus is really identical with any other, although they are associated with the same syndrome."
30. A 1992 memo produced by the United Nations Development Program emphasized this in an almost cynical way: "For most women, the main risk of infection by HIV is to be married. Every day 3,000 more women are infected and 500 infected women die." UNDP-Young Women: silence, susceptibility, and the HIV epidemic. (New York, 1992).
31. AIDS presents a more serious picture in Africa because of permanent infections among individuals (with or without AIDS), such as parasitoses that are also endemic (helminthiases). The environment thus plays an important role, as is shown by the history of Ethiopian immigrants in Israel. Following the "law of return," the whole Ethiopian Jewish community, the people who were called "Falasha" and are now known as "Beta Israel," migrated to Israel in two waves (1982–85 and 1991). Two percent of these immigrants were HIV-positive, infected with an HIV virus that was significantly different from the one infecting Israelis who were HIV-positive. Treatment of patients with worm parasites showed that the disappearance of this factor, which was connected with the Ethiopian environment, allowed these HIV-positive persons, who according to legend descended from the Queen of Sheba, to have a development like that of HIV-positive Israelis.
32. Gualde, N. *Ce que l'humanité doit à la femme*. Latresne: Le Bord de L'Eau, 2004, 44–45.
33. Tchak, S. *L'Afrique à l'épreuve du sida*. Paris: L'Harmattan, 2000.
34. Preston, R. *The Hot Zone*. New York: Random House, 1994.
35. In this work Dubos discusses the doctrine of "specific etiology," which assumes that a single pathogen corresponds to a single disease. This assertion bears the stamp of an ideological view of microorganisms that ignores environmental and psychological factors, and to which Dubos did not adhere. It was the way people in the Western world conceived of infectious diseases at that time.
36. Filoviruses of the family *Filoviridae* cause hemorrhagic fevers. Their genetic material is formed of RNA, and under an electron microscope it looks like filaments. The virus, which is vulnerable to common disinfectants, is nonetheless so dangerous that it can be handled in the laboratory only under conditions of maximum security.
37. The hospital in Yambuku had 120 beds for an area containing about six hundred thousand people. Care was provided by seventeen nuns, four of whom had nursing training, and a priest, without any certified doctor. The hospital treated three hundred to six hundred patients a day; they received a shot

containing, depending on the case, anti-malaria drugs, antibiotics, and, for pregnant women, vitamins. To give these shots, the nuns had five syringes. In Yambuku there were 361 infections and 318 deaths.

38. The number of articles in *Time* devoted to the problem of the transmission of disease, apart from the very special case of AIDS, shows how preoccupied North American media are with new epidemics that might reach their continent. The following table lists Ebola outbreaks up to 2005.

| Year | Ebola (subtype) | Country | Number of cases | Mortality rate (%) |
|---|---|---|---|---|
| 1976 | Zaïre | Democratic Republic of Congo | 318 | 88 |
| 1976 | Sudan | Sudan | 284 | 53 |
| 1976 | Sudan | Great Britain | 1 | 0 |
| 1979 | Sudan | Sudan | 34 | 65 |
| 1989 | Reston (monkey) | USA | 0 | 0 |
| 1990 | Reston | USA | 0 | 0 |
| 1992 | Reston | Italy | 0 | 0 |
| 1994 | Zaïre | Gabon | 44 | 63 |
| 1994 | Côte-d'Ivoire | Côte d'Ivoire | 1 | 0 |
| 1995 | Zaïre | Democratic Republic of Congo | 315 | 81 |
| 1996 | Zaïre | Gabon | 37 | 57 |
| 1996 | Zaïre | Gabon | 60 | 75 |
| 1996 | Zaïre | South Africa | 2 | 50 |
| 1996 | Reston | USA | 0 | 0 |
| 1996 | Reston | Philippines | 0 | 0 |
| 2000–2001 | Sudan | Uganda | 425 | 53 |
| 2001–2002 | Zaïre | Gabon, Democratic Republic of Congo | 122 | 79 |
| 2004 | Sudan | Sudan | 17 | 41 |
| 2005 | Zaïre | Congo | 11 | 82 |

39. The disease was given the name "Hantavirus pulmonary syndrome." This virus is now called the "Sin Nombre Virus" (SNV). The disorders to which it gives rise have been the subject of numerous reports. By September 1997, thirty-one cases had been found in New Mexico, twenty-two in Arizona, eleven in Utah, and ten in Colorado. The disease has appeared episodically in about half of the other American states but there has led to only a small number of cases.

40. The hantavirus pulmonary syndrome is a seasonal infectious disease, emerging in spring and summer, periods during which the rodents carrying the virus are the most numerous and the most active in their environment. Epidemiological studies have shown that the risk increases with the number of rodents. The virus responsible for the disease is not strictly speaking a new virus, and

only doubts regarding its nature at the time of its discovery caused it to be given different names, such as Sin Nombre Virus, Convict Creeks virus, and Muerto Canyon virus.

41. Before the cause of the disease was discovered, it was thought that it might be a new form of the flu virus, pulmonary plague, psittacosis (the latter transmitted by birds), Lassa fever, Marburg fever, other viruses, poisons, or pollutants such as mercury, nickel, and pesticides.
42. The tick grows from an egg that hatches a larva the size of a pinhead. This larva parasitizes small mammals (rodents). In winter, the larva becomes a nymph, still carrying the microorganisms from its host. The nymph can choose any warm-blooded animal as its host: raccoons, squirrels, birds, and humans, but it prefers deer (or lacking deer, a cow, horse, dog, cat, or human). It transmits to its host the agent that produces Lyme disease. The tick is an ectoparasite that lives on the blood of the parasitized individual.
43. At the beginning of the seventeenth century, New England was covered with beeches and cedars. The Indians customarily burned parts of the forest twice a year, but the damage remained minimal. At the beginning of the nineteenth century, New England's forests had been largely cut down [N. Jorgensen, *A Guide to New England's Landscape*. Barre, MA: Barre Publication, 1971]. Immense quantities of wood had been used for domestic or industrial ends (in the form of charcoal). In 1860 the New England forest was very limited, only 20 percent of its extent in 1992 after it had been "recuperated." The transfer of agricultural activity to the Great Plains helped make reforestation possible. The white-tailed deer (*Ododoileus virginianus*) served as a food source for carnivorous predators, especially during periods when the land was snow-covered. From 1600 to 1900, these deer were considered practically extinct. Thoreau described them as a local legend [H.D. Thoreau, *Walden, Or Life in the Woods*. Boston, Ticknor Fields, 1854]. From 1900 to 1960, the deer population underwent rapid growth, and the growth was even more explosive between 1960 and 1970.
44. The *Ixodes* ticks are abundant and diverse. In the United States, *I. dammini* and *I. capillaris* are found in the east, and *I. pacificus* in the west. In Europe, *I. ricinus* is found. The European parasite has left the countryside in order to attack residents of Paris, London, and Berlin.
45. A brief history of mad cow disease:

    April 1985: A rural veterinarian, Dr. Colin Whitaker, is called to a farm in Kent to examine a cow with abnormal behavior. This is considered the princeps case of BSE.

    November 1986: The first official diagnosis of bovine neurodegeneration with typical cerebral lesions.

    October 1987: BSE is officially recognized in the Veterinary Record in relation to cases in four herds.

    April 1988: After the appearance of BSE was correlated with cattle feed, experts concluded that "it is very unlikely that BSE will have any influence on human health."

July 1988: The use of animal carcasses is prohibited in the composition of feeds.

August 1988: The British government orders the slaughter of sick animals.

June 1989: The British government decides to prohibit the consumption of the bovine brain, thymus, and spleen—a regulation that was not to be applied until November.

March 1990: The European Community forbids importation into the continent of beef from Britain when the animals are more than six months old.

1993: One hundred thousand cases of BSE are reported.

1994: Six cases of the new form of Creutzfeld-Jakob disease are reported.

1995: Four additional cases of the new form of Creutzfeld-Jakob disease are reported. First deaths from the disease.

March 8, 1996: Dr. Rob Will of the BSE Surveillance Unit suggests that there may be a connection between mad cow disease and Creutzfeld-Jakob disease.

March 20, 1996: The British minister of Health states in the House of Commons that "The most likely explanation for the ten people who have died of Creutzfeld-Jakob disease was exposure to BSE before the 1989 prohibition."

—Prices of beef fall steeply.
—McDonald's and Burger King decide to stop using British beef.
—Britain's European partners within the Common Market decide to unilaterally prohibit British exports of beef and its derivatives.

March 27, 1996: The European Community prohibits the exportation of beef and its derivatives.

May 1996: The office of European veterinarians opposes a partial lifting of the prohibition.

November 1997: The problem of organ donations is raised. In a patient (Manon Hamilton) who died of lung cancer, eye tissue had been removed for use in three recipient patients. The donor's brain had turned out to be carrying Creutzfeld-Jakob disease.

December 2, 1997: It is officially recognized that spinal cord and bone marrow of animals infected with BSE are potentially contaminating.

December 3, 1997: Jack Cunningham, the British Minister of Agriculture, decides to prohibit the sale of beef with bones.

46. In 1983, George Balanchine died at the age of seventy-one. A naturalized American born in pre-Soviet Georgia, he was a brilliant choreographer (Diaghilev's former collaborator). In 1978, while living in New York, he began having trouble keeping his balance, an awkwardness that was unusual for someone who was normally so much in command of his body. This turned out

to be the first symptoms of Creutzfeld-Jakob disease, which was to produce, over the next five years, an irreversible neurological and mental involution that transformed an exceptional artist into a diminished being who had lost all autonomy, all rationality, and died in a lamentable state of physical and intellectual degeneration.

47. Ministère des Affaires Sociales de la Santé et de la Ville. Ministère Délégué à la Santé. "Precautions to be observed in surgical and anatomopathological environments in view of the risks of transmission of Creutzfeld-Jakob disease." Circular no. 45, July 12, 1994.
48. Scrapie is an ailment related to BSE. It is endemic in British sheep flocks.
49. Prusiner prions cause degenerative diseases of the central nervous system in the course of which cerebral gray matter becomes vacuolized. These diseases were named transmissible spongiform encephalopathies or slow-virus diseases. In fact, for a long time it was thought that they were caused by a virus called a "slow virus," one whose effects first manifested themselves years after contamination. The notion of a slow virus was connected with the work of Daniel C. Gajdusek, who in 1957 discovered the mode of transmission of Kuru, a culturally-related disease if ever there was one. Members of the Fore tribe in the highlands of Papua New Guinea practiced, as part of a mourning ritual, the consumption of the brains of their dead. This cannibalism provided a means of transmission of Kuru, which manifested itself in nervous disturbances and dementia (the Papuans called it "the laughing death"). Prion-related diseases strike humans and various animals (sheep, cattle, goats, cats, mink, elk, deer, etc.). Creutzfeld-Jakob disease is rare, and usually sporadic, affecting only one person per million worldwide. Nonetheless, there have been cases of hereditary Creutzfeld-Jakob disease, as there have of other prion-related hereditary diseases that are even rarer (Gerstmann-Sträussler-Schneiker syndrome and fatal family insomnia). It is remarkable that prions, which cause these various problems, are an abnormal form (PrP) of a normal protein that everyone produces. The appearance of the abnormal protein is thought to be due to a mutation in its gene, a mutation that leads to an intracerebral accumulation of the protein. When the disease (Kuru, mad cow) is transmitted, the abnormal prion protein ingested promotes the accumulation of prions in genetically predisposed individuals. This is in accord with the fact that the majority of the cases of Creutzfeld-Jakob disease connected with medical treatment affect patients with a homozygotic form of the mutation.
50. Stanley B. Prusiner won the Nobel Prize for Physiology and Medicine in 1997 for his work on prions. This ultimate recognition did not silence the most vociferous scientific critics, who do not accept the responsibility of prions for BSE and Creutzfeld-Jakob disease. In addition, there appeared in the popular press the story of young Clara Tomkins who, at the age of twenty-three, is in the final stages of Creutzfeld-Jakob disease. This young Englishwoman manifested the first symptoms about a year and half earlier, even though she

had been a vegetarian for more than a decade. [N. Gross, "Mad Cows and Humans." *Business Week,* December 22, 1997, 56.] However, the fact that one is a vegetarian does not preclude contact with the bovine proteins present in gelatin, lipstick, pharmaceuticals, and so forth.

51. A few of the "new" infectious agents identified since 1970 are as follows:

| Name of microbe | Year | Disease |
|---|---|---|
| Bacteria | | |
| *Legionnella pneumophila* | 1977 | Legionnaires' disease |
| *Campylobacter jejuni* | 1977 | Intestinal infections |
| *Staphylococcus aureus* | 1981 | Toxic shock syndrome |
| *Escherichia coli* 015:47 | 1982 | Colitis, hemorrhagic and uremic syndrome |
| *Borrelia burgdorferi* | 1982 | Lyme disease |
| *Helicobacter pylori* | 1983 | Stomach ulcers |
| *Vibrio cholerae* 0139 | 1992 | New cholera |
| Viruses | | |
| *Ebola* | 1977 | Hemorrhagic fever |
| *Hantavirus* (Asian) | 1977 | Hemorrhagic fever and renal syndrome |
| *Rotavirus* | 1979 | Infantile diarrhea |
| Human T-lymphotropic virus | 1980 | Leukemia |
| HIV | 1983 | AIDS |
| Human herpesvirus G | 1988 | Roseola |
| Hepatitis E virus | 1988 | Hepatitis |
| Hepatitis C virus | 1989 | Hepatitis |
| Guaranito virus | 1991 | Venezuelan hemorrhagic fever |
| Sin Nombre virus | 1993 | Hantavirus pulmonary syndrome |
| Sabia virus | 1994 | Brazilian hemorrhagic fever |
| Human herpesvirus 8 | 1995 | Kaposi's sarcoma |

| Name of microbe | Year | Disease |
|---|---|---|
| Parasites | | |
| *Cryptosporidium parvum* | 1976 | Diarrhea |
| *Enterocytozoon bieneusi* | 1985 | Diarrhea |
| *Cyclospora cayetanensis* | 1986 | Diarrhea |
| *Babesia* (new species) | 1991 | Atypical babesiosis |

52. For Gourou, the Wanade people are "a small group that has preserved the Bantu worldview in all its freshness." The Wanade, who are "woodcutters and proud of it," were forced to leave their homeland in Burundi, where there are no more trees, in order to settle in Kivu, in northwest Zaire.
53. Rift Valley fever, which is due to the virus of the same name, causes, in addition to fever, serious prostration, muscle pains, and headaches. Sometimes it takes the form of encephalitis or hemorrhagic fever.

54. Do we not speak of "sexual tourism" with reference to travel to certain countries?
55. Eastern equine encephalitis is due to an arbovirus present on the Atlantic coast of the Americas from the United States to Argentina. It affects horses and humans. The infection is very often asymptomatic, but it can also give rise to serious problems.
56. An investigation identified 367 different species in the ballast water of ships sailing between Japan and the port of Coos Bay, Oregon.
57. Dengue can cause mild problems or manifest itself as a serious hemorrhagic fever. The virus is transmitted by the *Aedes* mosquito (*A. aegypti, A. albopictus, A. polynesiensis*) and the *Stegomyia* mosquito. There is no interhuman transmission, no vaccine, and no specific treatment, but the illness confers immunity. Like all viruses carried by arthropods, *dengue virus* is an arbovirus.
58. We must distinguish here between population movements and travel. The former are involved in the emergence of diseases, while the latter is more often responsible for their spread.
59. The chief pathogens responsible for infectious diseases identified over the past two decades (after CDC) are as follows:

| Pathogen | Historical origin of the contamination |
|---|---|
| *Campylobacter jejuni* | chickens, milk |
| *Campylobacter fetus* subsp. *fetus* | |
| *Cryptosporidium cayetanensis* | |
| *Escherichia coli* | beef (hamburgers) |
| *Listeria monocytogenes* | |
| Virus Norwalk-like | oysters |

| Pathogen | Historical origin of the contamination |
|---|---|
| *Nitzchia pungens* | |
| *Salmonella enteritidis* | eggs |
| *Salmonella typhimurium* DT 104 | |
| *Vibrio cholerae* 01 | |
| *Vibrio vulnificus* | oysters |
| *Vibrio parahaemolyticus* | |
| *Yersinia enterocolitica* | pork |

## NOTES TO CHAPTER 3

1. The tuberculosis bacillus was discovered by Robert Koch in 1883; its scientific name is *Myobacterium tuberculosis*. This bacterium has the ability to survive within the white cells that phagocyte it in an attempt to defend the infected individual. The bacillus's ability to live inside phagocytes that have absorbed

it gives it an extraordinary pathogenic power. The genome of this redoubtable bacterium was recently sequenced, and genetic traits related to the specific pathogenic properties of Koch's bacillus were found.

2. Streptomycin and isoniazid are antitubercular bactericides. The advantage of isoniazid is that, unlike streptomycin, it acts within the cells where the bacterium has set up shop.
3. Phenomena of resistance to all antimicrobial molecules exist. This resistance can be natural, an intrinsic property of the germ with regard to the molecule, or it can be acquired, and concern only one strain of the microorganism. The selection of a strain is particularly common where the selective pressure exercised by antibiotics is strongest—usually in hospitals.
4. In 1983, Barbara McClintock won the Nobel Prize for her work showing that genes can change positions on the chromosomes of corn. The bits of mobile or "jumping" genetic material were called "transposons." Her work dates from the 1950s, but its importance long went unrecognized.
5. At the end of World War II, a Danish high school teacher who had pulmonary tuberculosis was "in contact" all year long with some three hundred students in a poorly ventilated classroom. At the beginning of the year, ninety-four of the students tested negative for tuberculosis. At the end of the year, twenty-four still tested negative; they had therefore not been "contaminated" by their teacher's tuberculosis bacteria. Twenty-nine tested positive; they had thus been infected by the bacillus without falling ill; their infection remained asymptomatic. Forty-one contracted the disease. After spending an academic year in a room with a tubercular teacher, 25.5 percent of the ninety-four students who tested negative remained negative and had thus escaped contamination. Almost 31 percent had a positive reaction without falling ill, and 43.6 percent got sick.
6. Governments that adopt the DOTS strategy agree to support tuberculosis monitoring, carry out bacteriological analyses of the sputum of individuals who have the disease or are suspected of having it, provide a standardized course of treatment lasting six to eight months, ensure a regular supply of suitable drugs, and collect medical observations in accord with certain standards.
7. Gerhard Henrik Armauer Hansen was a Norwegian physician who discovered the leprosy bacillus in 1873. Contagious victims of the disease expel the bacillus in their saliva and nasal secretions, as well as through skin lesions.
8. Exodus 4:6–8. "The Lord said to him [Moses] further, 'Put your hand into your bosom.' And he put his hand into his bosom; and when he took it out, behold, his hand was leprous, white as snow. Then God said, 'Put your hand back into your bosom.' So he put his hand back into his bosom; and when he took it out, behold, it was restored like the rest of his flesh. 'If they will not believe you,' God said, 'or heed the first sign, they may believe the latter sign.'"

Leviticus 13:9–11. "When a man is afflicted with leprosy, he shall be brought to the priest; and the priest shall make an examination, and if there is a white swelling on the skin, and there is quick raw flesh in the swelling, it is a chronic leprosy in the skin of his body, and the priest shall pronounce him unclean; he shall not shut him up, for he is unclean."

When Jesus cures the leper after the Sermon on the Mount, the accounts in Matthew, Mark, and Luke are virtually identical. Matthew 8:2–4: ". . . behold, a leper came to him and knelt before him, saying, 'Lord, if you will, you can make me clean.' And he stretched out his hand and touched him, saying, 'will; be clean.' And immediately his leprosy was cleansed. And Jesus said to him, 'See that you say nothing to any one; but go, show yourself to the priest, and offer the gift that Moses commanded, for a proof to the people.'"

9. "Lepers were often denounced by rumor, whereupon a court assigned physicians to submit an ad hoc report. Then the mayor's office sentenced the leper to sequestration, and informed the parish priest, who denounced the leper in his sermon. On the day set for the leper's 'expulsion from the world,' the priest blessed the drab gray mantle that leper was henceforth to wear. The leper retained only the mantle, a hat, gloves, clapper, belt, and knife. Soon the priests came to lead him in a procession to the church, followed by his relatives, friends, and neighbors gathered together in a kind of funeral cortege. The mass for the dead was sung, and the patient, isolated from his friends and relatives, listened to it with his face covered as if he were lying dead in his coffin. When the mass was over, the priest, after having given him the mantle, led him to the cemetery. There, he took three handfuls of earth and sprinkled them over the leper's head. 'My friend,' he said, 'you are dead to the world.' Then, pointing to the heavens, he urged him to resign himself." In Cabanes, G. *Moeurs intimes du passé. Les fléaux de l'humanité.* Albin Michel, 1955.
10. On January 30, 1998, on the occasion of the fiftieth anniversary of Gandhi's death, India announced that a new vaccine against leprosy, Leprovac, was to be put on sale. This vaccine was produced from a mycobacterium, *Mycobacterium w.*
11. In a single family in Surinam, the children sick with leprosy often shared the same tissue traits (HLA). [Table reproduced after De Vries, 82, 83].

| | Number of children having tissue traits in common | Number of leprous children who, in accord with the laws of chance, should have common traits |
|---|---|---|
| Tuberculoid leprosy | 188 | 139.6 |
| Lepromatous leprosy | 89 | 64.6 |
| Normal children | 128 | 125.8 |

12. The usual cause of malaria is inoculation of the protozoon through a mosquito bite. Only sixty of the four hundred species of the *Anopheles* mosquito are known to transmit malaria, and more precisely, only the females of these species. Malaria can also be contracted through blood transfusion, through contaminated syringes and needles, or during pregnancy (congenital malaria).
13. Laveran received the Nobel Prize for Medicine and Physiology in 1907 for the discovery of the parasite that causes malaria. Sir Ronald Ross showed in 1897 that the mosquito was the vector of malaria. In order to do so, he dissected mosquitoes for years before discovering the way in which the disease was transmitted. Ronald Ross won the Nobel Prize in 1902.
14. The tree is said to have been named "chinchona" because the viceroy's wife, Sra. Chinchón, benefited by treatment with it in 1638.
15. In 1820 Jean-Baptiste Caventou and Pierre Joseph Pelletier, both pharmacists, isolated alkaloids from the bark of the cinchona (quinquina), quinine and cinchonine. Quinine proved to be so effective that the botanist H. A. Weddell, in his *Natural History of Quinines,* worried that the forests would be destroyed in order to produce this drug. In 1852, the Dutch government established plantations of quinquina on the island of Java, and later the British did the same in India.
16. Hemoglobin is the basic protein that red cells contain and that serves to transport oxygen. In adult humans, the red cell contains adult hemoglobin (A). In the human fetus, red blood cells contain a different hemoglobin, fetal hemoglobin (F). In certain pathological circumstances, patients make hemoglobins that are abnormal in structure, like S hemoglobin, which causes sickle cell anemia. In other pathological circumstances, they produce fetal hemoglobin after their birth, a condition called thalassemia. Glucose-6-phosphate dehydrogenase is an erythocytic enzyme that participates in metabolizing glucose.
17. Living individuals are polymorphic, and this evident variety even within a species results from the fact that we have genes with several alleles, that is, genes that express traits having the same function but may differ. For example, genes for the ABO groups are responsible for the production of membrane structures that have the same role but whose nature may be A, B, or O. The ABO system is polymorphic. In France, the percentage of individuals in the groups A, B, AB, and O does not vary from one generation to another; it is balanced and has been stable since ancient times. The ABO system is polymorphic and balanced.
18. A normal adult has two alleles to produce adult hemoglobin. His genotype is thus HbA/HbA or more simply AA. One person who is ill with sickle-cell anemia may carry a single abnormal HbS allele; his genotype is AS, and he is heterozygotic. The genotype of another person who is suffering from the same disorder but who has two S-alleles will be SS; he is homozygotic for the

trait S. In this type of patient, the double dose of the S-trait usually results in death in infancy. The AS heterozygote presents major but not fatal clinical problems; he is protected against malaria.

19. "In 1950 the Duffy antigen was discovered in a multiply transfused hemophiliac, named Richard Duffy, in whose serum contained the first example of the Duffy antibody." At that time the great majority of Duffy-negative individuals were blacks from West Africa.
20. We now know that Duffy molecules on the surface of the cells serve as receptors for molecules that serve to transmit messages between cells: chemokines. Thus Duffy molecules are sometimes designated by the acronym DARC (Duffy antigen/Chemokine receptors).
21. HLA (human leukocyte antigens) designates the "traits," that is, the molecules that are present on human cells and that are involved in the rejection of grafts. When a graft is carried out, the compatibility between donors and receivers is more or less great in proportion to the number of HLA traits they share. In all individuals, HLA molecules also play a role in the process of immune response, namely the process that causes lymphocytes, the cells involved in immunity and the vigilant guardians of the organism's integrity, to "see" microorganisms and set in motion the processes of rejection.
22. In Vietnam, people in the hills and mountains live in houses built on stilts and keep their livestock on the ground floor. The women cook in the main room, constantly smoking up the house, which has no chimney. In these regions, the malaria vector, the *Anopheles minimus* mosquito, has a flight ceiling limited to three meters. The height of the houses and the smoke restricts attacks by mosquitoes on domestic animals and protects the hill people. The economic and political situation of the country led people living in the delta to emigrate to the mountains. These new arrivals built, according to their customs, low houses and cooked outside. These conditions facilitated their contamination by malaria, to the point that they thought they had been bewitched and left the area.
23. Resistant forms of malaria appeared along the border between Thailand and Cambodia. In the late 1960s, this area was in a state of anarchy, caught between the American forces and the Khmer Rouge. In 1975, the victorious Khmer Rouge emptied the cities of their inhabitants, leading to an exodus of thousands of refugees into Thailand. In Pol Pot's camps, prisoners were forced to carry out intensive irrigation projects, which produced sites for mosquitoes to reproduce. In addition, there were all the small reservoirs of water, from bomb craters to abandoned military vehicles, and tracks left by cars, carts and even the footprints of refugees. Finally, this region, which is rich in lodes of precious gems (rubies, emeralds), attracts miners hoping to make improbable fortunes among marshlands and streams in the tropical humidity in which mosquitoes proliferate.

24. *Streptococcus* is a bacterium that is often pathogenic for humans, especially when it belongs to the group *Streptococcus pyogenes.* There are streptococci responsible for pharyngitis, throat infections, scarlet fever, and impetigo, as well as later complications such as acute rheumatoid arthritis and Bright's disease (glomerulonephritis).
25. A septicemia or "blood poisoning" is a generalized infection of the bloodstream.
26. Group A streptococcus was extremely sensitive to penicillin (as well as to the most common other antibiotics). Ten thousand units of penicillin a day for four days were more than enough to cure an infection. Later on, in France the symptoms of scarlet fever were studied in medical books, because there were no more patients being treated for that disease. A previously common disease was at that time prevented by the use of antibiotics.
27. *Streptococcus pneumoniae,* or pneumococcus, is a bacterium that tends to settle in the lungs and that causes pneumonia, a pulmonary infection whose symptoms are the same in almost every case.
28. The WHO's statistics include respiratory diseases caused by viruses. It is estimated that around 2 percent of lung infections are viral in origin (measles, respiratory syncytial virus, flu). It is important to emphasize that the great majority of the childrens affected live in developing countries under extremely poor hygienic condition, and often suffer from nutritional deficiencies that have an impact on their immune system's functioning. The situation becomes positively grotesque when famines strike refugees crowded into camps where privation is combined with stress.
29. *Staphylococcus* is one of the microorganisms most pathogenic for humans. The bacterium can provoke abscesses (furuncles or boils) that may be superficial or deeper. Germs grafted onto certain organs can give rise to arthritis, osteomyelitis, endocardium of the heart, infections caused by food, and so forth. Among staphylococcus's peculiarities, let us note that it participates in the toxic shock syndrome mentioned earlier and is also frequently involved in nosocomial infections, particularly in the case of *Staphylococcus epidermis.*
30. See note 28.
31. Among the resistant bacteria responsible for diarrhea are *Shigella dysenteriae, Escherichia coli, Pseudomonas, Enterococcus faecium, Serratia marcescens,* as well as salmonellas and *Vibrio cholerae.* Among these microbial agents there are types resistant to many antibiotics, and in some cases only one antibiotic remains effective.
32. The mechanism that produces this phenomenon has not been fully explained, but it is at least partially related to the remains of antibiotic molecules in the animal feed and to the bacteria of the intestinal flora that produce vitamin $B_{12}$ (cyanocobalamin). Vitamin $B_{12}$ plays an essential role in hematopoiesis, that is, in the bone marrow's production of blood cells, and might be involved in

various trophic mechanisms. This molecule, which vertebrates cannot produce, is synthesized by bacteria in the digestive canal, but not by all these bacteria. By eliminating more easily the bacteria that do not produce vitamin $B_{12}$, animal feeds containing this vitamin facilitate the proliferation of the bacteria that do produce it. Ultimately, antibiotics are thought to produce animals with an excess of vitamins.

33. In Massachusetts, in 1991, an epidemic of diarrhea caused by *Escherichia coli* 0157:H7 broke out during the apple harvest. The bacterium was found in locally produced cider. The apples came from orchards in which the soil had been fertilized with animal manure that was probably infected with *E. coli* 0157:H7. The transmission of this bacterium from manure to humans is now well known.
34. When scientists seek to introduce a gene into a cell (for example, in corn), they associate the gene they are interested in with a second gene for resistance to antibiotics, which makes it possible, by adding antibiotics to the medium when the cells are cultured in the laboratory, to select the resistant forms that have therefore incorporated the two genes. Thus, genetically modified plants have a gene for resistance to antibiotics. These genes have been found in the digestive systems and in the blood of animals that have eaten feeds made from genetically modified plants.
35. *Herpesvirus hominis* or human herpes simplex virus (HSV-1 and HSV-2) produces infections of the skin and mucous membranes as well as of nerve cells and, more rarely, the internal organs. The virus can attack the genitals and lead to general symptoms (fever, fatigue, etc.). If the affected individual has no immune deficiencies, the disease tends to regress, but the infection is very often recurrent, always producing painful symptoms. When the individual has an immune deficiency, the infection can be very serious. It is treated with local antivirals with or without acyclovir, which is the preferred injectable drug used to fight the disease.
36. The artificial expansion of structures related to irrigation (dams, reservoirs) has encouraged diseases caused by parasites such as malaria and schistosomiasis. In both cases, it does so by facilitating the multiplication of the intermediate host (mosquitoes or snails). This was seen in Mali and in Zanzibar during a period when rice cultivation was being expanded. In Ethiopia, the incidence of malaria has doubled as a result of the migration of a nonimmune population to an "economic development" zone in an area where malaria is endemic. There are numerous examples of increases in the frequency of malaria connected with agriculture and deforestation.
37. The Onchocerciasis Control Program (OCP) launched in West Africa was a great success and was officially terminated in December 2002. In 1995 a broader program was launched for the whole of Africa, the African Program of Onchocerciasis Control (APOC).

38. Regarding the emergence of AIDS, many silly things have been said and written regarding the responsibility of one group or another. Thus, the epidemic was seen as a result of divine punishment, an intergalactic virus, a CIA conspiracy, sexual relations between Africans and apes, and so on. So far as bizarre explanations go, our contemporaries have invented nothing new: at the time when syphilis was epidemic, the disease was called, depending on the speaker, the American, French, or Neapolitan disease. Its origin was said to be divine or related to bestiality. Ozanam, citing Van Helmont, thinks that the disease resulted from a sexual relationship between a man and a mare that had a runny nose. Linder had (already) proposed relations with apes, and Manard, relations with a leper, as the cause.
39. Here, chosen as examples, is some of the information given on the WHO's Web site for the week of January 5–11, 1998:

    January 5: confirmation of sixteen cases of flu caused by an A virus ($H_5N_1$) in Hong Kong, four of which were fatal.

    January 6: outbreak of cholera in the Democratic Republic of Congo (eight hundred cases with fifty-four deaths since December 18).

    Outbreak of Rift Valley fever in Kenya: the disease strikes humans and sheep, goats, cattle, and camels. Three hundred deaths are reported in a poorly accessible area in the Garissa district in northeast Kenya.

    January 7: the epidemic of cholera continues in Tanzania. Since the end of January 1997, there have been 35,591 cases and 2,025 deaths.

    January 9: the WHO is concerned about the advance of cholera in eastern and southern Africa.

    January 11: outbreak of cholera in Chile.

## NOTES TO CHAPTER 4

1. The main infectious agents encountered among human beings and the diseases they cause are as follows:

| Agents | Diseases |
|---|---|
| Bacteria | |
| *Salmonella* | Typhoid fever |
| *Pneumococcus* | Pneumonia |
| *Myobacterium tuberculosis* (Koch bacillus) | Tuberculosis |
| *Myobacterium leprae* | Leprosy |
| *Streptococcus b hemolyticae* | Throat infections, scarlet fever, erysipelas |
| *Treponema pallidum* | Syphilis |
| *Bordetella pertussis* | Whooping cough |
| *Neisseria gonorrheae* | Gonorrhea |
| *Staphylococcus aureus* | Abcesses, inflammations |
| *Yersinia pestis* | Plague |

| Agents | Diseases |
|---|---|
| Virus | |
| Myxovirus | Mumps |
| Morbillivirus | Measles |
| Myxovirus | Influenza |
| Rhinovirus | Colds, pneumonia |
| Enterovirus | Diarrhea |
| Variola | Smallpox |
| HIV1 and HIV2 | AIDS |
| | |
| Parasites | |
| *Candida albicans* | Thrush |
| *Pneumocystis carinii* | Pneumonia |
| Plasmodium | Malaria |
| Trypanosoma | Sleeping sickness |

2. The antibodies produced by humans are called immunoglobulins, and are of several types: M (or IgM), G (IgG), A (IgA), D (IgD), and E (IgE).
3. In 1956, Jean Dausset, who was then an assistant physician for Paris hospitals and head of a laboratory at the National Center for Blood Transfusion (Centre National de Transfusion Sanguine), published a work on immunohematology in which he described, and illustrated with a photograph, how leukocytes (white blood cells) agglutinate. This approach, which Dausset himself described as "the heroic time of leuco-agglutination," made possible the discovery of the first antigenic leukocyte (MHC-A2) recognized as an antigen of histocompatibility, that is, a product involved in graft rejection. HLA is the international acronym for "human leucocyte antigen"; it is also called the major histocompatibility complex or MHC. For Dausset, as for other physicians confronted by the problems of incompatibilities in transfusions and graft rejection, the MHC system was the biological entity complicating treatment of certain patients. These physicians saw themselves as concerned with humans' rejection of human tissues.
4. The HLA genes present on their chromosome (no. 6) are not always associated by pure chance. For example, the frequency with which we should find on a single chromosome the gene HLA-B8 and the gene HLA-DR3 is the theoretical frequency, calculated (Fc) = the product of the frequency of each gene in the population studied; thus, for Europe, Fc(B8-DR3) = F(B8) × F(DR3), or 0.9 % × 10 % = 0.09 %. Studies of populations have shown that the theory is not always verified by observation; in a given population the frequency of the observed links (Fo) may differ from the theoretical frequency. For B8 and DR3, the observed frequency is 7 percent, therefore superior to Fc (0.09 percent). The difference Δ between Fo − Fc is called a

linkage disequilibrium; here $\Delta = 6.1$. The linkage disequilibrium shows that in humans there are favored connections among certain genes; therefore, the distribution of characteristics is not wholly aleatory and subject to the calculus of probability.

5. The low rate of HLA alleles in Amerindian populations are as follows:

| Population studied | Number of HLA alleles present |
|---|---|
| Africans | 40 |
| Europeans | 37 |
| Asians | 34 |
| Amerindians (northern) | 17 |
| Amerindians (southern) | 10 |

6. Measles is still a fatal disease, especially in Africa. Worldwide, one million children die each year from this disease. These deaths correlate with the relative absence of vaccination against measles. Almost 90 percent of young Europeans and Americans are vaccinated; in Asia only 75 percent of the children are vaccinated, and in Africa, hardly 50 percent.
7. The mortality associated with measles is not linked solely to physiological poverty or genetic characteristics but also depends on the way in which the disease is transmitted. Viral infections are often more serious when they are transmitted within family groups, especially if extensive interactions among the children of a poor family lead to prolonged, repeated transmission of the virus between those already contaminated and those not yet contaminated. For unknown reasons, the seriousness of the transmission between relatives is increased when the children involved are of different sexes.
8. In the history of humanity, three major epidemiological transitions are generally recognized. The first occurred ten thousand years ago when a nomadic way of life was abandoned in favor of a sedentary one and the creation of agriculture, that is, with the breakup of an ecosystem that (already) favored the emergence of new diseases. The second transition was that involving the hope of eradicating infectious diseases. The third is the one we are currently experiencing, with emergences that are erasing the second epidemiological transition.
9. Polio, like many infectious diseases, can benefit from social disorders and from poverty. At the end of 1996, the disorder reappeared in Albania, infecting 120 children and leading to sixteen deaths. In the current situation, the country can only count on the help of foreign institutions.
10. Differences, according to human groups, in the appearance of poliomyelitis among residents of Casablanca in the 1950s are as follows:

| | Europeans | Moroccans |
|---|---|---|
| Residents | 125,000 | 530,000 |
| Patients with polio | 117 | 25 |
| Cases (percentage of cases per year per 100,000 residents) | 13.7% | 0.7% |
| Cases per age group in 1953 | | |
| 0–2 | 9 | 0 |
| 2–9 | 15 | 2 |
| 10–39 | 5 | 0 |

11. The first of these theories, that of Lamarck is called the theory of the inheritance of acquired characteristics, that is, of characteristics acquired for the purpose of adaptation. This theory of evolution, that of "necessity without chance," no longer has many adherents among scientists. The second theory, Darwin's, is that of natural selection, according to which only the individuals best adapted to their environment survive. Modernized and transposed to the molecular level, it has given rise to the neo-Darwinian or synthetic theory of evolution, that of "chance and necessity," as Monod famously put it. The third theory, that of Kimura is called "neutralist" or "stochastic," and is opposed in many respects to the synthetic hypothesis. For Kimura, most mutations are neutral, and within a given population the evolution of mutated genes may follow the vagaries of genetic drift; in this case, it is a matter of "chance without necessity." For the neo-Darwinians, on the contrary, mutated genes are more or less subject to favorable or unfavorable selective pressures that affect the persistence or disappearance of the characteristics coded by these genes.
12. Microbes are not the only factor in the evolution of the MHC; in mice, the choice of reproductive partner seems to be connected with the MHC, since the animals prefer to mate with individuals having different MHCs, differences that are discerned by the sense of smell. Among the descendents of these couples, we find more heterozygotes than would theoretically be expected; it is as if the individuals behaved in such a way as to engender offspring with the most polymorphic MHC possible, whence the choice of a "different" partner, as if this were an ancestral and biological form of the incest taboo. The latter, by prohibiting mating between related individuals, limits the probability of engendering descendents that are homozygotic so far as their HLA alleles are concerned and thereby limits HLA monomorphism.
13. The principal associations between HLA and infectious diseases have to do with parasitic infections (malaria, scabies, leishmaniasis, kala-azar), viral diseases (hepatitis B, hepatitis C), bacterial infections, including those due to mycobacteria (leprosy) and Klebsiella infections.

14. Favorable and unfavorable associations between ABO groups and infectious diseases are as follows:

| Diseases | Favorable groups | Unfavorable groups |
|---|---|---|
| Typhoid | O | A, B, AB |
| Tuberculoid leprosy | O | |
| Lepromatous leprosy | A | O |
| Syphilis | O | A, B, AB |
| Influenza | O | A |
| Hepatitis | A | O |
| Cholera | O | A |
| Primary tuberculosis | O | A, B, AB |

15. The myxomatosis virus, of Brazilian origin, was introduced into Australia in 1950. In 1964, 99 percent of the wild rabbits were dead, but a tiny minority survived. Over the years, several types of virus replaced, one after the other, the original agent of the disease.
16. Resemblances between microorganisms and human tissues, and diseases associated with this phenomenon are as follows:

| Microorganisms | Tissues |
|---|---|
| Streptococcus | Heart, kidney |
| Klebsiella | HLA-B27 |
| *Escherichia coli* | Colon |
| Trypanosoma | Nerves, heart |

## NOTES TO CHAPTER 5

1. In mechanics and mathematics, the term "chaos" refers to apparently random or unpredictable behavior in systems governed by deterministic laws.
2. The Viking probe was sent to Mars in March 1976 to take samples and make analyses of the planet's soil in the hope of finding evidence of past or present life there. The result was negative, as Lovelock had predicted, basing his prediction on the composition of Mars's atmosphere, which lacks methane. For Lovelock, the presence of free methane in Earth's atmosphere proves that this gas (which combines easily) results from a living process (particularly anaerobic bacteria).
3. The "biotope" is the geophysical environment, whereas "biocenosis" refers to the interactions among all living beings inhabiting a biotope. Together, the biotope and biocenosis constitute an ecosystem.

4. Margulis, L. and G. Hinkle, "The Biota and Gaia," in L. Margulis and D. Sagan, *Slanted Truths: Essays on Gaia, Symbiosis and Evolution*. New York: Copernicus Books, 1997.
5. Laplace wrote: "We may regard the present state of the universe as the effect of its past and the cause of its future. An intellect which at any given moment knew all of the forces that animate nature and the mutual positions of the beings that compose it, if this intellect were vast enough to submit the data to analysis, could condense into a single formula the movement of the greatest bodies of the universe and that of the lightest atom; for such an intellect nothing could be uncertain and the future just like the past would be present before its eyes." This "intellect" has become known as "Laplace's Demon." However, as Edgar Morin has written, ". . . uncertainty cannot be dissipated, because no one, not even Laplace's demon, could have an objective point of view from which he could discern the future of the universe, and thus diagnose its past."
6. Henri Poincaré (1854–1912) was a mathematician who studied mechanics, physics, and astronomy. He examined the "three-body problem," which concerns the movements of three masses subjected only to their mutual attraction, the context of his work on celestial mechanics. Responding to a challenge issued by the king of Sweden, a challenge intended to show that the solar system described by Newton was dynamically unstable, Poincaré, without providing a definitive proof, laid the foundations of the notion of chaos.
7. Fractals are geometric forms that repeat themselves on different scales. The geometry involved is characterized by invariance of scale, that is, its synthesis is found on all scales. In other words, one part of the object reproduces the whole on a smaller scale.
8. It is impossible to discuss chaos without mentioning the Santa Fe Institute (SFI), a pluridisciplinary research center founded in 1984 whose members explore complex phenomena. They study the origin of life on Earth, the evolution of the stock market, the universal laws of biology, the self-organization of systems, animal behavior, artificial life, and so forth. One of the most eminent figures at SFI is Murray Gell-Mann, winner of the Nobel Prize for physics and the author of a book on complex systems entitled "*The Quark and the Jaguar.*"
9. "Most of my colleagues believe that life emerged simple and became complex. They picture nude RNA molecules replicating and replicating and eventually stumbling on and assembling all the complicated chemical machinery we find in a living cell. Most of my colleagues also believe that life is utterly dependent on the molecular logic of template replication, the A-T, G-C Watson-Crick pairing . . . I hold a renegade view: life is not shackled to the magic of template replication, but based on a deeper logic . . . life is a natural property of complex chemical systems . . . when the number of different

kinds of molecules in a chemical soup passes a certain threshold, a self-sustaining network of reactions—an autocatalytic metabolism—will suddenly appear. Life emerged, I suggest, not simple, but complex and whole, and has remained complex and whole ever since—not because of a mysterious élan vital, but thanks to the simple profound transformation of dead molecules, into an organization by which each molecule's formation is catalyzed by some other molecule in the organization. The secret of life, the wellspring of reproduction, is not to be found in the beauty of Watson-Crick pairing, but in the achievement of collective catalytic closure. The roots are deeper than the double helix and are based in chemistry itself. So, in another sense, life—complex, whole, emergent is simple after all, a natural out-growth of the world in which we live. The origin of life on the planet is more the result of setting in motion chemical catalytic systems and complexes in a state of non-equilibrium that are in a situation of a Prigogine dissipative system, than the result of processes in which the duplication of DNA and/or RNA plays a major role."

10. Kurt Gödel was an Austrian logician who became an American citizen in the 1940s. He formulated "the most profound conceptual result obtained by humanity in the course of the twentieth century." He explained that there are assertions, even when they are formulated very precisely, whose truth or falsity cannot be demonstrated. For mathematicians, the demonstration of the incompleteness theorem was a serious blow.
11. A nearly exhaustive overview of the notion of chaos is provided by the works of Ilya Prigogine, in particular by *Order Out of Chaos* (written in collaboration with Isabelle Stengers). See also his "La thermodynamique de la vie" [pp. 223–50 in 178], *Les lois du chaos,* and more recently *La fin des certitudes.* On chaos theory in biology see especially Jacques Tonnelat's two-volume work published under the titles *Thermodynamique et biologie* and *L'ordre issue du hasard.*
12. Vittorio Volterra (1860–1940) was an Italian mathematician. The father-in-law of Umberto D'Ancona, a biologist, he was inspired by the latter to undertake a study of the population of fish in the Adriatic. After two months of study, Volterra proposed his mathematical model. Alfred J. Lokta, an American biologist, produced analogous models independently. The Lokta-Volterra model describes the interactions between two species in an ecosystem. One species is a predator, the other its prey, and the model is thus represented by two equations, that of the predator and that of the prey.
13. Measles owes its name to the spots it produces on the skin, but the human measles virus also has a pulmonary tropism and temporarily alters immune defenses.
14. Genetic studies of the virus infecting dolphins in Denmark have shown that it is different from other morbilliviruses affecting dogs (canine distemper virus) and seals (phocine distemper virus), as well as from human measles virus.

15. High concentrations of PCBs (polychlorinated biphenyls) have been found in the fat of sick Mediterranean dolphins.
16. Autopsies performed on the Mediterranean dolphins in oceanographic laboratories in Barcelona, Valencia, and Murcia showed that they had lost weight and presented, above all, signs of pneumonia. The virus was present in the bronchial and bronchiolar tissues. The brain was also damaged. The dolphins' encephalitis could be connected with other pathogens such as toxoplasmosis, and this suggests that the animals had a substantial immune deficiency.
17. One gram of human feces contains about one billion viruses, many of which are not infectious. A liter of human "sewer water" contains 100,000 infectious viruses, most of which are resistant to chlorine. What then should we say when the excrements of entire continents such as Africa and Asia flow into the ocean via streams, rivers, and other watercourses?
18. "Herd immunity" can be defined as a group's resistance to attack by an infectious disease to which a large number of individuals eventually become immune. The term seems to have been first used in 1929 by Topley and Wilson, who used it to emphasize the difference between the immunity of a group and that of an individual.
19. C. Combes, *Les associations du vivant. L'art d'être parasite.* (Paris: Flammarion, 2001) 22.
20. Kraus, A. "Diverse host responses and outcome following simian immunodeficiency virus SIVmac239 infection in sooty mangabeys and rhesus macaques." *Journal of Virology.* 1998; 72: 9597–9611.
21. Palacios, E. "Parallel evolution of CCR-null phenotypes in humans and in a natural host of simian immunodeficiency virus." *Current Biology* 8 (1998): 943–946.
22. Nowak M. A. "The mathematical biology of human infections." *Conservation Ecology* 3 (1999): 12–18.
23. Morell, V. "How the malaria parasite manipulates its hosts." *Science,* 1997: 278; 223.
24. Every human newborn has an immunity in the form of antibodies passively acquired in utero via the placenta. The protection the mother gives the baby lasts three to six months, so that the "immunological period of coming into the world" the one in which the human newborn has to grapple alone with his microbial environment, is deferred with respect to the official birth date.
25. Combes, C. *Interactions durables.* Paris: Masson, 1995.
26. Moles, A. A. *Les sciences de l'imprecis.* Paris: Seuil, 1995.
27. In his book Jean-Pierre Dupuy frequently refers to the works of Edgar Morin and some less often to Douglas Hofstadter's *Gödel, Escher, Bach: An Eternal Golden Braid.* These authors discuss the self-organization of complex phenomena, but we must also emphasize that in volume 2 of his *Méthode* Morin writes: "I call classical science any scientific procedure that is governed by

a paradigm of simplification (. . .) The paradigm of simplification operates through reduction (of the complex to the simple, of the molar to the elementary), rejection (of the aleatory, of disorder, of the singular, of the individual), and disjunction (between objects and their environment, between subject and object)." We know how acerbic Morin can be with regard to traditional scientific procedures.

28. Factors involved in lymphomas

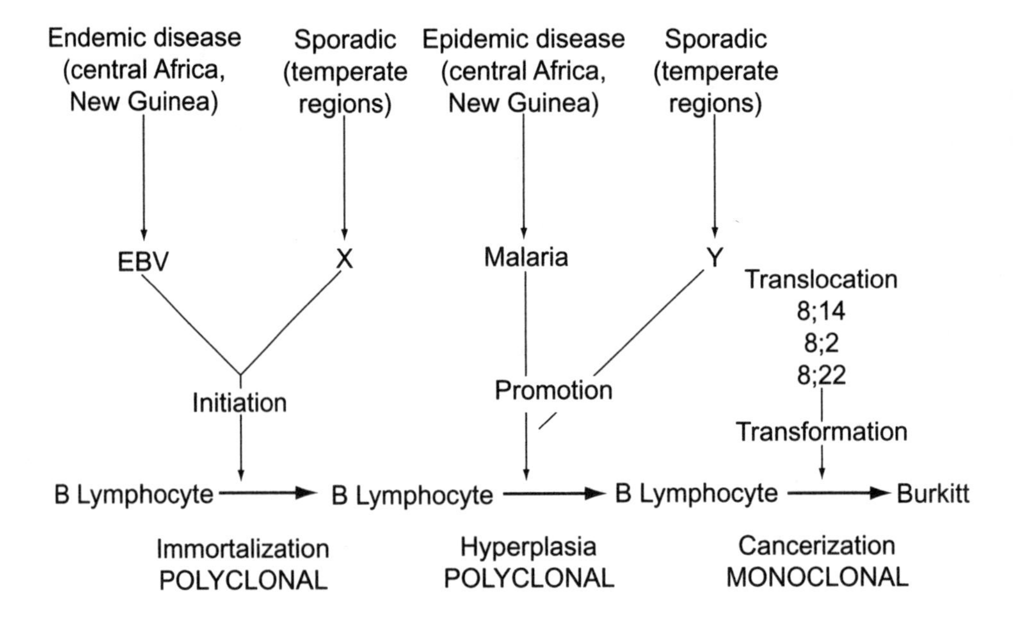


The schema above summarizes, so far as is possible, the complexity of the viral (Epstein-Barr virus), parasitological (malaria), chromosomal (translocations), and ethnic factors involved in the occurrence of Burkitt's sarcoma. In regions where it is endemic, the environmental factors are more or less well-known; in Europe, the disease differs in its causes, which are less well-known (X, Y, . . . ?). The latter is "another" Burkitt's lymphoma, although in both cases the cancerous cells are immune system cells (B lymphocytes).

29. Cohen, J. E. "Human population grows up." *Scientific American*. September 2005: 26–33.
30. Gualde, N. et al. "Immunologie du vieillissement." *Revue de Médecine Interne* 4 (1983): 231–238.
31. Yoshikawa, T. T. "Aging and infectious diseases: state of the art." *Gerontology* 30 (1984): 275–278.

32. Felser, J. M. et al. "Infectious diseases and aging: immunologic perspectives." *Journal of the American Geriatrics Society* 31 (1983): 802–807.
33. Cook, J. M. et al. "Alterations in the human immune response to the hepatitis B vaccine among the elderly." *Cellular Immunology* 89 (1987): 109–113.
34. Pechère, J.-Cl. "Le défi microbiologique des pneumonies bactériennes du sujet âgé acquises en dehors de l'hôpital." *Médecine et Hygiène* 51 (1993): 2492–2495.
35. Roberts-Thomson, I. C. et al. "Ageing, immune response, and mortality." *Lancet* 2 (1974): 368–370.
36. Martin, G. M. "Interactions of aging and environmental agents: the gerontological perspective." *Progress in Clinical and Biological Research* 228 (1987): 25–80.
37. Ruffié, J. *De la biologie à la culture.* Paris: Flammarion, 1976.
38. Meredith, P. J. et al. "Autoimmunity, histocompatibility, and aging." *Mechanisms of Ageing and Development* 9 (1979): 61–77.
39. Partridge, L. et al. "Optimality, mutation and the evolution of ageing." *Nature* 362 (1993): 305–311.
40. Martin, G. M. "Interactions of aging and environmental agents: the gerontological perspective." *Progress in Clinical and Biological Research* 228 (1987): 25–80.
41. Olshansky, J. B. Carnes, and C. Cassel. "Le vieillissement de l'espèce humaine." *Pour la Science* 188 (1993): 32–39.
42. Hawkes, K. "Grandmothers and the evolution of human longevity." *American Journal of Human Biology* 15 (2003): 380–400.
43. Rogers, A. R. "Why menopause?" *Evolutionary Ecology* 7 (1993): 406–420.
44. Beise, J. et al. "A multiple event history analysis of the effects of grandmothers on child mortality in a historical German population. Krummhörn, 1720–1874." *Demographic Research* 7 (2002): 469–497.
45. Moulin, A. M. *Le dernier langage de la Médicine. Histoire de l'immunologie de Pasteur au sida.* Paris: Presses Universitaires de France, 1911.
46. "The WHO has used the Internet to strengthen its global networks responsible for monitoring this or that disease, such as the WHO data bank on resistance to antimicrobial agents, FluNet, RABNET, and Global Salm-Surv, which enable national centers and collaborating centers worldwide to exchange information on pharmacoresistance, flu, rabies, and salmonellosis." Report of the Secretariat of the Fifty-fourth World Health Assembly.
47. http://www.who.int/disease.outbreak-news/
48. http://www.who.int/wer
49. Drucker, J. *Les détectives de la santé. Virus, toxiques: enquêtes sur les nouveaux risques.* Paris: Nil, 2002, 15.

50. Monitoring of AIDS with the help of screening draws on four sources of information: blood banks, anonymous, free screening centers; supervisory physicians (1 percent of general practitioners), a network of four hundred laboratories (the RENA VI network). Investigations of seroprevalence in which the subjects remain anonymous have been carried out on pregnant women in the Paris region and in the Provence-Alpes-Côte d'Azur region of southeastern France, in clinics when screening for venereal diseases is being done, and in health care centers where semiannual "on a given day" investigations are made.
51. The annual incidence of cases of AIDS is decreasing across Europe, from the east to the west. In 1994 it was, per million inhabitants, 191 in Spain, 100.3 in Italy, 98.4 in France, and 98 in Switzerland.
52. These are stationary satellites or ones that have polar orbits and are operated by the National Ocean and Atmospheric Administration (NOAA, United States) or the European consortium EUMETSAT (Meteosat).
53. Consider SARS and avian flu. The history of epidemics is endless, and the diversity and size of the microbial world are such that anything is possible on the stage occupied by the agents of epidemics.
54. "Vaccines of the future." In *The Dana Sourcebook of Immunology,* ed. Dan Gordon. (Washington, D.C.: Dana Press, 2005) 51.
55. "Therapeutic vaccines." Ibid., 52.
56. Bacteriophages produce lysines specific to bacteria and are capable of developing antiresistance systems.
57. The peptides strategy is based on the ability of certain fragments of cyclical peptides to form nanotubes that can punch fatal "holes" in the cell walls of even the most resistant bacteria.
58. Gualde, N. *Immunité et humanité. Essai d'immunologie des populations.* (Paris: L'harmattan, 1994.) 187 and 190.
59. I had the honor and the pleasure of regularly talking with Professor Rosen at Harvard in the late 1970s. He was an extraordinarily competent, cultured man who was open to the world; he was as well-informed about truffle omelets in Perigord as about lymphocytes. It was with sorrow that I learned of his death just as he was about to receive the first AAI (American Association of Immunologists) Dana Foundation Award in Human Immunology.
60. Charles A. Janeway, Jr. *Immune Surveillance of the Brain in Health and Disease.* Yale University School of Medicine. Dana Grant Programs 2004.
61. MacDonald, A. "Dana Sponsored Neuroethics Conference Begins to Define Issues, Conflict in Emerging Fields." *Dana Alliance Member News.* Vol. VI, No. 3. (June/July 2002): 1–3.
62. See Morin H., La méthode. 5. *L'humanité de l'humanité. L'identité humaine.* Paris: Seuil, 2001.
63. See, for example, the reports regularly published by *Immunology in the News* (Dana Press's quarterly publication).

1. In order to improve detection of outbreaks of these pathogens, particularly in the food supply, France has established the Institut de veille sanitaire (InVS; "Institute for Public Health Monitoring"). This state institution, operating under the aegis of the minister of health, constantly monitors public health, studies the environment, and analyzes national lifestyles. The InVS collaborates with other agencies that monitor public health and coordinates its activities with those groups.
2. The heroes of this semi-fictional book, set at the beginning of the twentieth century, are very poor Lithuanian immigrants working in a meat-packing plant. In chapter 14, we read that the sausages were made of "meat kept in piles in storerooms where water was dripping from leaks in the roof and thousands of rats were running around. In these storerooms it was too dark to see well, but a man could pass his hand over these piles of meat and brush off an armful of rat droppings. These rats were a nuisance, and the packers set out poisoned bread; the rats died, and then rats, bread, and meat all went into the hoppers to be ground up." It is easy to understand why this novel (which sold over 100,000 copies) alarmed Americans regarding the poor quality of their food, which they were discovering for the first time. On the other hand, there was rather less concern for the plight of the unfortunate workers.
3. UNAIDS, 20 Avenue Appia, 1211 Geneva 27, Switzerland. HTTP://www.unaids.org
4. Anthrax is caused by a bacterium, *Bacillus anthracis*. The disease is usually vocation-related, for the bacterium is found in sheep's wool and hair of bovines. Anthrax decimates herds, and may infect those who handle wool and leather. The disease, which has been known since antiquity, was mentioned by the Roman poet Virgil, who noted that it was transmitted from sheep to humans. In Greek, "anthrax" means "coal," and the disease was so named because of the dark color of infected animals' blood and of skin lesions in humans. Anthrax can be cutaneous, in which case it is not very serious, or pulmonary, when it is very serious. The bacterium, which has been adapting to its environment for millions of years, is remarkably resistant.
5. As of October 2001, thirty-two countries had not yet signed the convention: Algeria, Andorra, Angola, Antigua and Barbados, Azerbaijan, Cameroon, Chad, Comoros, the Cook Islands, Djibouti, Eritrea, Guinea, the Vatican, Israel, Kazakhstan, Kiribati, Kyrgyzstan, the Marshall Islands, Mauretania, Micronesia, Moldavia, Mozambique, Namibia, Nauru, Niue, Palau, Samoa, Sudan, Tadzhikistan, Trinidad and Tobago, and Vietnam.
6. In an article published in *Le Monde* (October 6, 2001), Jeremy Rifkin, the president of the Foundation on Economic Trends, listed the countries concerned: South Africa, Bulgaria, China, North Korea, South Korea, Cuba, Egypt, India, Iran, Iraq, Israel, Laos, Libya, Russia, Syria, Taiwan, and Vietnam.

7. The FBI is supposed to have had proof of an attack by means of contaminated letters as early as September 18, 2001. In early October, two senators also received such letters.
8. Concerning smallpox, observers noted that when the USSR was taking a vigorous part in the battle against the disease—providing, for example, large doses of vaccine—the Soviet empire was producing massive quantities of the smallpox virus for military uses. What was going on? Consummate Machiavellianism? Bureaucratic inconsistency?
9. American health authorities have opted for "ring vaccination." Briefly, this means that in the event of an outbreak of smallpox caused by terrorists, patients would be isolated, and anyone who had been in contact with them would be vaccinated.

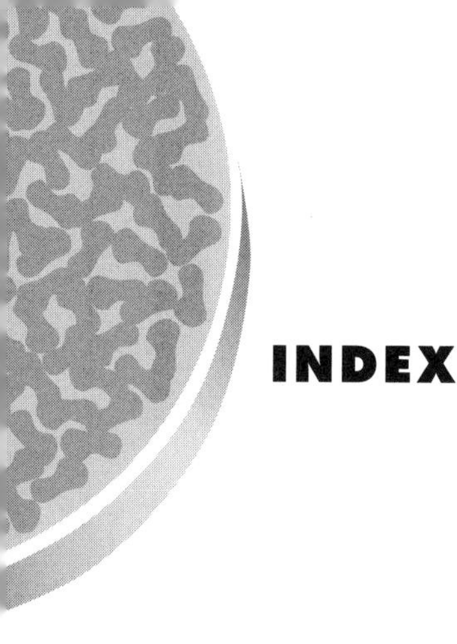

# INDEX


A
ABC approach to AIDS pandemic, 36
abiotic factors, 19, 27, 92, 132. *See also* climate conditions
ABO system, 105, 116, 178*n*17, 186*n*14
acquired immunity, 103
acquired immunodeficiency syndrome. *See* AIDS
adaptive system, mammals' immunity as, xi
*Aedes aegypti* mosquito, 17, 18, 59, 163*n*26
*Aedes africanus* and *A. simpsoni* mosquitos, 18
*Aedes albopictus* mosquito, 60
*Aedes camptorhyncus* mosquito, 55
*Aedes* mosquitos, 57, 175*n*57
Afghanistan, malaria in, 82
Africa
  biological resistance to HIV, 37, 38
  cholera, 182*n*39
  current epidemics, 28
  HIV/AIDS and women, 37, 39–40, 167*n*22, 169*n*30, 192*n*50
  HIV and AIDS, 31, 36, 151
  and HLA alleles, 107
  Lassa fever, 45
  leprosy, 75
  malaria, 76, 77–78, 82
  measles fatalities, 184*n*6
  prostitution in, 32, 37, 40, 59–60, 151, 166*n*10, 167*n*20
  *See also specific African countries*
African Americans and AIDS, 67
African green monkey *(Cercopithecus aethiops sabaeus),* 31, 44
African Program of Onchocerciasis Control (APOC), 181*n*37
Afro-Americans' resistance to HIV, 38
*Ages of Gaia, The* (Lovelock), xii
aging, continuous nature of, 140
aging process, 140–141
agriculture
  antibiotics fed to livestock, 87–88
  bacteriological warfare based on, 157, 158
  deforestation and, ix, 46, 56
  and emergent plagues, ix, 46, 57–58, 93–94
  irrigation systems and, 57, 181*n*36
  microbe-supportive practices, 21
  shift from nomadic life to, 184*n*8
AIDS (acquired immunodeficiency syndrome)
  ABC approach to, 36
  about, viii, 29–30

AIDS (*continued*)
area and poverty-level differentiation, 34, 36, 39–40
biological resistance to, 37–39
bizarre explanations for, 182*n*38
CD4 cells and immunity, 37–38
choices about, 150–153
current statistics on, 150–153
decrease in incidence of, 192*n*51
denial and fatalistic response to, 36, 63, 152–153, 166–167*n*12
emergence of, 30–31, 165*n*2–3
epidemiology of, 30–37
Europeans' immunity to, 38, 168*n*27
future of, 39–40
and HLA traits, 37, 167*n*22
incidence of, 67
initial identification, 30, 165*n*1
monitoring, 192*n*50
monitoring of, 145
opportunistic infections with, 30, 35, 39–40, 165*n*1
polio vaccine origin theory, 151–152
pregnant women with, 37, 39, 167*n*22, 192*n*50
treatments for, 33–34, 37, 167*n*14, 167*n*19
tri-therapy treatment for, 35–36, 37, 167*n*14
*See also* HIV
airline flight personnel and tuberculosis, 72
Alaskan Eskimos, 107–108
alastrim. *See* smallpox
Alibek, Ken (was Kanatjan Alibekov), 154
alleles
Amerindian populations, 184*n*5
CCR5-32 HIV-resistant allele, 39, 168*n*27
for hemoglobin production, 178*n*18
MHC and HLA alleles, 107
Amazon region, 82, 112
Amerindian populations
biological characteristics, 20, 38, 108–112, 184*n*5
origin of, 109
Spanish colonization and, 2–3, 20
*Analytic Theory of Probability* (Laplace), 127
anarchical situations
breakup of family group, 166*n*10
drug use, 33, 34, 35, 63–64, 67, 166*n*11
*Anopheles* mosquito
and apes, 112–113
discovery of relationship to malaria, 178*n*13
fifth century B.C. Egyptians' awareness of, 76–77
genome modification hypothesis, 146–147
insecticide resistance of, ix–x, 82–83
and protozoon plasmodium parasite's life cycle, 79–80
species and sex of, 178*n*12
Vietnamese solution to, 179*n*22
*See also* malaria
anthrax
post-9/11 terrorist attacks, 155–156, 194*n*7
Russian bacteriological warfare plans, 154–155
in WWI, 157
anthrax bacterium *(Bacillus anthracis)*, 193*n*4
anthropic modifications of the environment, 149–158
about, 58–61

antibiotics fed to livestock, 87–88, 180*n*32
demography, outbreaks, and, 61–64
emergence of diseases related to, ix
feeding sheep to cows, 53
and flu virus mutagenesis, 68
genetically modified plants, 88, 181*n*34
and Legionnaires' disease, 48–49
and Lyme disease, 49–51
and microbe modifications, 21
and outbreaks, 68
overpopulation and poor hygiene, 8–9, 69
permanent threat modulated by, 120–121
and resistance, 92–95
social and ecological disorder, 12
spontaneous disappearance of infectious diseases in spite of, 91–92
treatments for diseases, 120–121
*See also* deforestation
anthropocultural feedback loop, 146
antibiotics
in genetically modified plants, 181*n*34
hope for the future, vii–viii, 184*n*8
hopes and realities, 83–88
for leprosy, 75
in meat and eggs, 87–88, 180*n*32
overuse results in *Shigella* diarrhea, 87
plague bacilli with resistance to, 12
resistance to, 70–71, 85–86, 87–88
unconventional, 146
antibodies, 183*n*2
antiDuffy antibody, 179*n*19–20
antigenic leukocytes, 183*n*3
anti-malarial measures, 82–83
antimicrobial peptides, 146
antiresistance systems of bacteriophages, 192*n*56
antitubercular bactericides, 176*n*2
antivirals, 88–89
apes. *See* primates
APOC (African Program of Onchocerciasis Control), 181*n*37
*Apodemus agrarius* mouse, 57–58
*Arenaviridae* virus, ix, 45–46, 58
"Are some people immune to AIDS?" *(Time)*, 37
Argentina
hemorrhagic fever, ix, 58
Junin virus, ix, 46, 58
Argentine hemorrhagic fever, ix
Arizona, 46–48, 54–55, 170*n*39
Asia
cholera, 14
diphtheria, 21–22
HIV and AIDS, 33–34, 36, 151
Japanese encephalitis, 58
Korean hemorrhagic fever, 57–58
lack of biological resistance to HIV, 38
malaria, 76, 82
plague epidemics, 7
prostitution, 33, 34, 35, 36, 63, 151
*See also specific Asian countries*
Athens, Greece, plague in, 24
Australia
HIV and AIDS, 34
myxomatosis, 22, 116, 163*n*29
and myxoma virus, 8, 22, 117, 163*n*29, 186*n*15
Ross River virus, 55
autoimmune diseases, 138–139
Averroes, 22
avian flu, 192*n*53
avian viruses, 16–17
Azerbaijan, diphtheria in, 21–22

## B

bacillus Calmette-Guérin (BCG) vaccine, 72
bacteria
- and deforestation, ix
- exchange of genetic material, 86
- gleaning antibiotics from, 146
- and human ecology, 86–88
- "new" infectious agents discovered, 174*n*51
- as "wild beasts" of the future, 18–19
- *See also specific bacterium*

bacterial meningitis, 107
bacteriological warfare, 153–155
bacteriophages, 146, 192*n*56
Bak, Per, 128
Balachine, George, 172*n*46
Balkhash, Lake (central Asia), 7–8
*Balto* (film), 26
Bangladesh, 2, 14, 162n20
Barbier, Marguerite, 162–163n23
BCG (bacillus Calmette-Guérin) vaccine, 70, 72
Bible
- on leprosy, 74, 176–177*n*8
- on plague, 6, 7, 160n9

biocenosis, 56, 57–58, 94, 124, 186*n*3
*Biohazard* (Alibek), 154
Biological and Toxic Weapons Convention (1972), 154, 155, 193*n*5
biological cycle of protozoon plasmodium, 79–80
biological immunity, x–xi, 80–81. *See also entries beginning with* "immunity"
biological markers
- of individuals in the immunology of populations, 104–105, 116, 138
- and infectious diseases, 105–106
- of resistance or susceptibility to HIV, 37–38
- and vaccinations, 106–108

biological predisposition
- to cholera, 14, 162n22
- to malaria, 80
- O blood type and cholera, 14, 162n22

biological resistance
- to HIV, 37–39
- to leprosy, 75–76
- *See also* immunity of populations

biology of epidemics, 19–21. *See also* biosociocultural aspects of epidemics
biomathematical models, 134
biosociocultural aspects of epidemics
- about, 18–22, 19, 96–97
- addressing all the issues continuously, 27–28
- AIDS, 39
- cultural aspects of epidemics, x–xi, 26–27, 81–82
- *See also* anarchical situations; immunity of individuals; immunity of populations; social aspects of epidemics; *entries beginning with* "bio"

*Biota and Gaia, The* (Margulis), 126
bioterrorism, 155–158
biotope, 186*n*3
Black Death, 7. *See also* plague
blood flukes, 106
blood poisoning, 180*n*25
blood transfusions, 66
blood type relationship to cholera, 14, 162n22
Blue Death. *See* cholera
B lymphocytes, 103–104, 110, 114, 141, 179*n*21, 190*n*28
Boccacio, 6, 160n7
Bolivia, Machupo virus, ix
Bolivian hemorrhagic fever, 46

books. *See* writings
Borclet, J., 163*n*27
Bordeaux, France, 8–9, 161*n*12, 161*n*16
Bordet, Jules, 116
Borrel, Amédée, 137
*Borrelia burgdorferi* bacterium, 49–50, 51, 174*n*51
Botswana, HIV and AIDS in, 36
Boudard, Alphonse, 69
bovine spongiform encephalopathy (BSE), x, 51–53, 60, 149–150, 157, 171*n*45
Braudel, Fernand, 6, 160*n*8
Brazil
  blood flukes, 106
  hantavirus, 48
  leprosy, 75
  malaria, 82
  Oropouche virus, 31, 62, 165–166*n*4
  schistosomiasis, 106
Bredig's heart, 127, 128
Breughel the Elder, 6
Brinckmann, 94, 95
BSE (bovine spongiform encephalopathy), x, 51–53, 60, 149–150, 157, 171*n*45
*Bunyaviridae* (hantavirus), 47
Burkitt's sarcoma and lymphoma, 137, 190*n*28
Burma, 74, 82
Burnet, MacFarlane, 138
Burundi massacre and AIDS pandemic, 166*n*12
Bush, George W., 157
*Business Week,* 34

C
*Calomys callosus* mouse, ix, 46, 47, 58
Cambodia, 82, 179*n*23
Canada, 34, 48
cancer, 136–138, 190*n*28
Carroll, James, 17
Casablanca polio outbreak, 184*n*10
Castonguay, Daniel, 153
cathelicidines "caths," 146
Catholicism, 11, 26, 42
Caventou, Jean-Baptiste, 77, 178*n*15
CCD (Cold-Cloud Duration), 145
CCR5-32 HIV-resistance allele, 39, 168*n*27
CCR5 (chemokine receptor 5), 38, 133, 168*n*24
CD4 (complex of differentiation 4), 38, 110
CDC (Centers for Disease Control), 30, 70–71
cellular receptors for HIV, 38, 39, 110, 133, 168*n*24, 168*n*27
Centers for Disease Control (CDC), 30, 70–71
Central America, yellow fever in, 17
central nervous system, x, 173*n*49
*Cercopithecus aethiops sabaeus* (African green monkey), 31, 44
cerebral immuno-monitoring, 147
cerebral system and immune system, 147
chaos
  about, 126–128, 186*n*1
  deterministic chaos, 127–128, 136
  infectious, 147–148
  MetaGaia and chaos model, 136
  three-body problem, 187*n*6
chaos theory, 123, 126–131, 132, 134, 135, 187–188*n*9–11
Chastel, Claude, 5
chemokine receptor 5 (CCR5), 38, 133, 168*n*24
Chile, cholera in, 182*n*39
chimpanzee *(Pan troglodytes troglodytes),* 31–32
China
  contributory agricultural practices, 21

China (*continued*)
HIV and AIDS, 35, 36, 151
husbandry practices, 68
inoculation method, 4
Japan's bacteriological warfare on, 153–154
Korean hemorrhagic fever, 57–58
cholera
about, 12–13
biological predisposition, 14
and *El Niño,* 14, 54–55, 161*n*19
relationship to water, 13–14, 22, 161*n*18
cholera bacterium *(Vibrio cholerae)*
about, 12–13, 161*n*17
blood type relationship, 14, 162*n*22
transported in ships' ballasts, 14, 21, 60, 175*n*56
water as supportive environment, 13–14, 22, 161*n*18
cholera vibrio, modifications of, 20–21
chromosomal relationship to lymphomas, 190*n*28
cinhona tree, 77, 178*n*14–15
climate conditions
and chaos theory, 126–127
*El Niño,* 14, 54–55, 161*n*19
and emergent plagues, 22, 54–55
and overpopulation of *Calomys* mice, 47
and *Ptychodiscus brevis* alga, 131–132
coevolution of mammals and microbes
about, xi, 132–133
cultural partner in, 118
influenza virus and humans, 20
malaria and, 133
microbes' ability to hide in less virulent forms, 22–23
myxomatosis and rabbits, 22, 163*n*29
plague bacillus and humans, 7–8, 10
*See also* genetics
Cold-Cloud Duration (CCD), 145
collective catalytic closure theory, 187–188*n*9
Colombia, leprosy in, 75
Colorado, hantavirus in, 46–48, 54–55, 170*n*39
Combes, Claude, 132, 134, 135
*Coming Plague, The* (Garrett), 18–19, 42, 87
complex of differentiation 4 (CD4), 38, 110
*Conquista,* 2–3, 23–24, 108–109, 121
Costa Rica, 55–56
Côte d'Ivoire Ebola, 41, 170*n*38
Côte d'Ivoire Ebola epidemic, 56–57
Creutzfeld-Jakob disease, x, 51–53, 60, 173*n*49–50
Crusades, 59
Cuba, yellow fever in, 17
*Culex* mosquitos, 57, 119
cultural aspects of epidemics, 26–27, 81–82. *See also* biosociocultural aspects of epidemics; immunity of populations
cultural immunity and biological immunity, x–xi
culture. *See* spirituality and culture
Curtis, Tom, 151–152
*Cynomolgus* monkeys, 43–44
cytokines, 85, 103

D

Daisyworld, 124–125
DARC (Duffy antigen/Chemokine receptors), 179*n*20
Darwin, 185*n*11
Dausset, Jean, 104, 183*n*3

DDT (dichlorodiphenyltrichloroethane), viii, ix, 46, 89, 121
deaths. *See* mortality rates
Debré, Robert, 105
*Decameron* (Boccacio), 160*n*7
deer mouse *(Peromyscus maniculatus)*, 47, 55
deer tick *(Ixodes scapularis)* parasite, 49–50, 53
Defoe, Daniel, 160*n*10
deforestation
  emergent diseases from, 55–57, 94, 96
  and Lyme disease, 50–51, 171*n*43
  and Machupo virus, ix, 46
  and malaria, 82–83, 181*n*36
  and Oropouche virus, 31, 62, 165–166*n*4
  *See also* forest ecology
dehydration, from cholera, 12–13
Delumeau, Jean, 152–153
Democratic Republic of Congo
  and AIDS pandemic in Africa, 59–60, 166*n*12
  breakup of family group, effects of, 166*n*10
  cholera, 182*n*39
  Ebola virus treatment in Yambuku hospital, 42, 169*n*37
  monkeypox, 160*n*6
  Zaire Ebola, 41, 42, 169*n*37
demography, 23–24, 61–64, 75, 78, 142. *See also* mortality rates
dengue virus, 60, 62, 158, 175*n*57
denial and fatalistic responses to AIDS, 36, 63, 152–153, 166–167*n*12
Dessein, H., 106
deterministic chaos, 127–128, 136
diarrhea, from resistant bacteria, 180*n*31
dichlorodiphenyltrichloroethane (DDT), viii, ix, 46, 89, 121
diffusion. *See* pathogen diffusion
diphtheria, 21–22
directly observed therapy—short course (DOTS strategy), 73, 176*n*6
diseases
  infectious agents related to, 182–183*n*1
  and opportunistic infections, 30, 35, 39–40, 165*n*1
  resemblance between human tissues and microorganisms, 186*n*16
  *See also* infectious diseases; pathogen diffusion; *specific diseases*
dissemination. *See* pathogen diffusion
dissipative structures, 127
dissipative system, Prigogine-type, 187–188*n*9
Djibouti, cholera, 28
DNA
  biological and cultural, x–xi, xi
  and cancer, 136
  and gene therapy, 146
  in humans, 101, 127
  of resistant bacteria, 71, 86
  in "vaccines of the future," 146
doctrine of specific etiology, 169*n*35
dolphins, 129–131, 188*n*14, 189*n*15, 189*n*16
domestic animals and infectious diseases, 58
domestic terrorism, 156
DOTS strategy (directly observed therapy—short course), 73, 176*n*6
drug use epidemic and HIV, 33, 34, 35, 63–64, 67, 166*n*11
Dubos, René, 41, 67, 169*n*35

Duffy, Richard, 179*n*19
Duffy antigen/Chemokine receptors (DARC), 179*n*20
Duffy blood group, 80, 116, 179*n*19–20
Dumont, René, 55
Dupuy, Jean-Pierre, 135, 189*n*20
Durban, South Africa, pneumonia in, 86

E

*E. coli* bacterium, 64, 94–95, 150, 174*n*51, 181*n*33
Easter Island smallpox epidemic, 3
Eastern equine encephalitis, 60, 175*n*55
Ebola, 29, 170*n*38
Ebola virus, 41–44, 169*n*36
ecological aspects of epidemics
  about, 19, 22
  and emergent plagues, 54–58
  and hemorrhagic fevers, 46
  natural role of microorganisms, 100
  overpopulation of *Calomys* mice, ix, 47, 58
  unbalanced biocenosis, 56, 57–58, 94, 124, 186*n*3
  *See also* biosociocultural aspects of epidemics; climate conditions; deforestation
ecological niche and malaria, 119–120
ecology of populations and chaos theory, 129
economic impact of plague, 11
ecosystem, 186*n*3, 188*n*12
ectoparasite, 171*n*42
eggs and *Salmonella enteriditis* bacterium, 65
Egypt, 69, 106, 113
Ehrlich, Paul and Anna, 57
elderly persons
  aging process, 140–141
  grandmother theory, 141–142
  immunity of, 66, 118, 139–140, 141–142
*El Niño,* 14, 54–55, 145, 161*n*19
encephalitis in dolphins, 189*n*16
England
  cholera pandemic, 13
  mad cow disease timeline, 51–52, 171*n*45
  plague, 11
  sweating sickness epidemic, 91–92, 106
*Enquête sur un nouveau paradigme* (Dupuy), 135
environment, man as polluter and victim, x
epidemics
  about, 18–19, 27–28, 96–97
  chaos theory for study of, 134
  consequences of, 23–28
  difficulty of forming single theoretical model, 133–134
  human and animal compared, 131–132
  plague, 7
  smallpox, 1–3
  WHO information on, 28, 182*n*39, 191*n*46
  *See also* fear of epidemics; pathogen diffusion
epidemiological transition, 112
epidemiological transitions in human history, 184*n*8
epidemiology
  of AIDS, 30–37
  considering biological and social aspects, 19
  and influenza virus, 17
Epstein–Barr virus, 190*n*28
eradication of microbes, belief in, vii–viii
erythocytic enzyme, 178*n*16

*Escherichia coli* bacterium, 64, 94–95, 150, 174*n*51, 181*n*33
Ethiopia, malaria in nonimmune population of, 181*n*36
ethnic factors in lymphomas, 190*n*28
Europe
  Caucasians' biological resistance to HIV, 38
  cholera as disease of poor, 12, 13
  decrease in incidence of AIDS, 192*n*51
  HIV and AIDS, 34, 35, 36
  Lyme disease, 50
  persecution of Jews and lepers, 9, 25, 161*n*13, 164*n*33–34
  plague, 7, 8–10
  polio, 113
  smallpox epidemics, 2
  tuberculosis, 73
  *See also specific countries*
evolution
  biological and culture, x–xii
  of immunity of populations, 113–118
  of infectious diseases, 7–8
  of influenza virus A, 163*n*25
  of microbes, 67–68
  of plague bacillus, 8
  theories on, 185*n*11

F
FDA (Food and Drug Administration), 88
fear of epidemics
  about, 24–25, 164*n*31
  accusations of plague-spreading, 25, 164*n*35
  leprosy, 74
  persecution of Jews and lepers resulting from, 9, 25, 161*n*13, 164*n*33–34
  of smallpox, 1–2, 159*n*1
feedback loops, 19, 27, 54, 93–95, 147
filarial disease, 90
films, 26, 29, 165*n*36
filoviruses *(Filoviridae),* 41–42, 45–46, 169*n*36
Finlay, Carlos, 17, 163*n*26
Fischler, Claude, 149, 150
flagellants, 26
flavivirus (yellow fever), 17–18
fleas *(Xenopsylla cheopsis),* 6–7
"flesh-eating bacterium," 84
flu. *See* influenza
Food and Drug Administration (FDA), 88
food industry
  about, 64–65
  dangers of contaminated food, 149–150
  and *E. coli,* 94
  FoodNet program, 144–145
  Sinclair's novel on, 150, 193*n*2
FoodNet program, 144–145
forest ecology
  and emergent plagues, 55–57
  and Lyme disease, 50, 171*n*42, 171*n*44
Fracastor, 69
fractals, 127, 187*n*7
France
  ABO system stability, 178*n*17
  cholera, 12, 13
  Institut de veille sanitaire, 193*n*1
  persecution of Jews, 164*n*33–34
  plague epidemics, 7, 8–10, 161*n*11, 161*n*12, 161*n*15
  prion-detection systems, 150
  smallpox epidemics, 2

G
Gaia
  chaos or, 128–131, 135–136
  MetaGaia and chaos model, 136

Gaia (*continued*)
symbol of harmony, xi
virtues of, 125–126
Gaia hypothesis, 123–126, 128–129
Gajdusek, Daniel C., 173*n*49
Galvani, A. P., 168*n*27
Gambia, malaria in, 145
Ganges delta, cholera in, 12
Garrett, Laurie, 18–19, 42, 87
Gates, Bill, 36
"gay phase" of AIDS, 30
Gell-Mann, Murray, xi, 187*n*8
gene therapy, 146–147
genetically modified plants, 88, 181*n*34
genetic engineering, 29
genetic predisposition, 80–81, 173*n*49, 177*n*11, 178*n*18
genetics
of Amerindian populations, 108–112, 184*n*5
bacteria sharing genes for resistance, 86
CCR5 cellular receptors, 38–39, 133, 168*n*24, 168*n*27
CD4 cellular receptors, 38, 110
cholera and, 12, 14
of dolphins, 188*n*14
immunity as hereditary characteristic, 22, 163*n*28
of influenza virus, 16–17, 163*n*25
microbe mutations, 20, 39, 67–68, 133, 169*n*29
microbes' gene modification capability, 20–21
monomorphism of Amerindians, 108–112
theories on gene mutations, 125
transposons, 176*n*4
*See also* biological markers; DNA; HLA (Human Leukocyte Antigens)
genome, of tuberculosis bacillus, 175*n*1
geometry, 187*n*7
Georgia (Asia), diphtheria in, 21–22
Germany, 25–26, 161*n*13
*Germs: Biological Weapons and America's Secret War* (Miller), 154
Gheerbrant, Alain, 112
glanders, 157
Gleick, James, 123, 127
globalization, 58–62, 90
Global Malaria Eradication Program (WHO), 77
global monitoring, 143–144, 144, 162*n*20, 192*n*50
glucose-6-phosphate dehydrogenase, 79–81, 178*n*16
Gödel, Kurt, 128, 188*n*10
Gourou, Pierre, 55, 174*n*52
grandmother theory, 141–142
Greece, plague in, 24
Greek mythology, xi
*Grippe de 1918 á 1919 dans un service de l'hôpital Sant-Antoine, La* (Barbier), 162–163*n*23
Grmek, Mirko, 30, 56
Gross, Ludwig, 137
Group A streptococcus, 83–84
Group B streptococcus, 83
Grubler, Arnulf, 61

H

*Haemophilus influenzae* bacterium, vaccination for, 107–108
Haiti, HIV and AIDS in, 33
Haldane, J. B. S., 7–8, 80, 116
Hansen, Gerhard Henrik Armauer, 176*n*7
Hansen's bacillus *(Mycobacterium leprae)*, 73, 75
Hansen's disease, 73–76, 177*n*10
Hantaan virus, 57–58

hantavirus *(Bunyaviridae)*, 47
hantavirus pulmonary syndrome or Sin Nombre virus (SNV), 46–48, 54–55, 170*n*39–40
Hausen, Harald Zur, 137, 138
Hawaii, avian malaria in, 119–120
Hawkes, Kristen, 142
hematopoiesis, 180*n*32
hemoglobin, 79–81, 117, 178*n*16, 178*n*18
*Hemophilus influenzae* bacterium, 68
hemorrhagic viral fevers
  about, 46
  from agrarian reform in Bolivia, ix
  Argentine, ix
  and biological warfare, 155
  Bolivian, 46
  dengue, 175*n*57
  Ebola, 29, 41–44, 169*n*36, 170*n*38
  and filoviruses, 169*n*36
  identified agents of, 174*n*51
  Korean, 57–58
  Lassa fever, 45–46
  Marburg fever, ix, 44, 158
  monitoring outbreaks, 144
  Rift Valley fever, 57–58, 68, 174*n*53, 182*n*39
  yellow fever, 17–18, 22–23, 28, 59, 62, 163*n*26
Henderson, Henzel, x–xi
Henson, Jim, 84
hepatitis B, 66, 106
hepatitis C and blood transfusions, 66
herd immunity, 132, 189*n*18
hereditary immunity, 22, 163*n*28
*Herpesvirus hominis,* 89, 137, 174*n*51, 181*n*35
heterosexual relations and HIV, 33, 34, 36–37, 39–40
*Histoire du Sida* (Grmek), 30
HIV (human immunodeficiency virus)
  anomalies to contraction of, 39
  biological resistance to, 37–38
  from blood transfusions, 66
  drug use and, 33, 34, 35, 63–64, 67, 166*n*11
  effect of mobility on, 59–60
  heterosexual relations and spread of, 33, 34, 36–37, 39
  HIV-negative women delivering HIV-positive infants, 39
  mutations of, 39, 169*n*29
  origin of, 31, 151–152
  resistance to AZT, 89
  safe sex, 33, 34, 37, 39, 167*n*20
  social inequalities as infection co-factor, 34, 35
  statistics on, 151
  and tuberculosis, 71–72
  women and, 37, 39–40, 167*n*22, 169*n*30, 192*n*50
  *See also* AIDS
HLA (human leukocyte antigens) system
  and ABO microbes, 116
  about, 104–105
  and antigens, 110–112, 114–115
  associations of, 183*n*4
  and autoimmune diseases, 139
  biological markers and, 106–107
  defenses against malaria and, 81
  discovery of, 183*n*3
  gene therapy for, 146
  HIV and HLA traits, 37–38
  and infectious diseases, 185*n*13
  polymorphism of, 115–116, 119
  as portion of population's immune potential, 119
  and skin graft compatibility, 80–81, 104, 179*n*21
  targeting for gene therapy, 146
Hoechst, 44
homosexuality and AIDS, 30
Hong Kong, influenza in, 182*n*39

hoof-and-mouth disease, 157
Hooker sea lions *(Neophoca hookeri)*, 129–130
Hooper, Edward, 151
*Hospital, a Patient's Biography, The* (Boudard), 69
hospital in Yambuku, syringe reuse in, 66
hospitals, contracting *Staphylococcus* in, 84–85
*Hot Zone, The* (Preston), 29
HSV-1 and HSV-2 (human herpes simplex virus), 89, 137, 174*n*51, 181*n*35
human activities. *See* anthropic modifications of the environment
human beings
  biological evolution, xi, 113–118
  bizarre explanations for epidemics, 182*n*38
  and Gaia hypothesis, 125
  immunity of newborns, 133, 189*n*19
  menopause, 141–142
  resistance to safe sex, 33, 34
human feces, 21, 189*n*17
human herpes simplex virus (HSV-1 and HSV-2), 89, 137, 174*n*51, 181*n*35
human leukocyte antigens (HLA) system. *See* HLA
human population, 139
hygiene
  and cholera, 13
  food industry's lack of, 150, 193*n*2
  and hantavirus, 47
  and measles, 184*n*7
  overpopulation and, 8–9, 20, 42, 62, 69
  and plague, 9–10
  and respiratory diseases in children, 180*n*28
  and tuberculosis, 72
  *See also* social disorders

I
iatrogenesis, 52, 65–67
IDU (injection drug use), 67
IGAS (Intensive Group A Streptococcus), 84
immune deficiencies of seals and dolphins, 130–131
immune system
  adaptability of, 135
  autoimmune diseases and, 138–139
  cerebral system and, 147
  coevolution of microbes and, xi
  disease treatments and, 66
  of elderly persons, 66, 118, 139–140, 141–142
  failure to respond to prions, 52–53
  and HLA molecules, 179*n*21
  partial immunity from, 93
  stress to, 8–9
  *See also* coevolution of mammals and microbes; HLA; MHC
immunity
  of elderly, 140
  of human newborns, 133, 189*n*19
  lymphocytes and, 103–104, 110, 114, 141, 179*n*21, 190*n*28
  and malaria, 82–83
  natural versus acquired, 103
  skin and, 102–104
  to yellow fever, 18
immunity of individuals, 99–108
  biological markers and immunology of populations, 104–105
    ABO system, 105
  biological markers and infectious diseases, 105–106
  biological markers and vaccinations, 106–108

immunity of individuals, 101–104
See also HLA system
immunity of populations, 99–100, 108–121
about, 118, 159*n*3
and CCR5-32 HIV-resistance allele, 39, 168*n*27
evolution of, 113–118
factors in, 133–134
herd immunity, 132, 189*n*18
and human activities, 112–113
and polymorphism, 80, 119–120, 178*n*17
immunoglobulins, 104, 183*n*2
immunohematology, 183*n*3
immunologic responders, 10
incompleteness theorem, 128, 188*n*10
India
cholera, 14
HIV and AIDS, 34, 36
leprosy, 74
typhus 2000 B.C., 56
Indonesia, leprosy in, 74
*Infection et immunité* (Borclet), 163*n*27
infectious agents, diseases related to, 182–183*n*1
infectious chaos, 147–148
infectious diseases
and ABO groups, 186*n*14
doctrine of specific etiology, 169*n*35
feedback loops, 19, 27, 54, 93–95, 147
and HLA, 185*n*13
human migration and, 61
mortality rates, 143
pathogens responsible for, 175*n*59
*See also specific diseases*
influenza, 14–15, 16–17, 20, 163*n*24. *See also* Spanish flu pandemic
influenza virus
A and B designations, 16
modifications to, 20–21, 68, 117–118
writings on, 29
injection drug use (IDU), 67
inoculation, origin of, 3, 4
Institut de veille sanitaire (InVS–France), 193*n*1
intellect, 187*n*5
Intensive Group A Streptococcus (IGAS), 84
*Interactions durables* (Combs), 134
interleukins, 103
International AIDS Conference, 152
InVS–France (Institut de veille sanitaire), 193*n*1
irrigation and disease, 181*n*36
isoniazid, 176*n*2
Israel, parasitoses and AIDS, 169*n*31
Italy, typhus in, 121
ivermectin, 90–91
*Ixodes dammini, I. capillaris, I. pacificus, and I. ricinus* ticks, 49–50, 53, 171*n*44
*Ixodes scapularis* (deer tick) parasite, 49–50, 53
*Ixodes* ticks, 49–50, 53, 171*n*44

J
Jahrling, B., 159*n*4
Janeway, Charles A., Jr., 147
Japan, 90, 106, 153–154, 156
Jenner, Edward, 4
Jews, persecution of, as scapegoats for plague, 9, 25, 161*n*13, 164*n*33–34
*Journal of the Plague Year, A* (Defoe), 160*n*10
Juliana's disease. *See* AIDS
*Jungle, The* (Sinclair), 150, 193*n*2
Junin virus *(Arenaviridae)*, ix, 46, 58
juvenile rheumatoid arthritis, 49

K
Kaposi's sarcoma, 30, 137
Karlen, Arno, 58
Kauffman, Stuart, 128
Kenya, 28, 182*n*39
Kimura, Naoko, 125, 185*n*11
Klein, Jan, 115
Koch, Robert, 175*n*1
Koch's bacillus, 69–72, 175*n*1
Koprowski, Hilary, 152
Korean hemorrhagic fever, 57–58
Koster, Fred, 46
Kuru transmission in Papua New Guinea, 173*n*49

L
Ladurie, Emile Le Roy, 22
Lake Baikal seals *(Phoca sibirica),* 130, 131
Lamarck, Jean-Baptiste, 185*n*11
Landsteiner, Karl, 105
Laos, malaria in, 82
Laplace, Pierre-Simon, 127, 187*n*5
Laplace's Demon, 187*n*5
Lassa fever *(Arenaviridae),* 45–46
Lassa virus, ix, 44, 45–46, 158
"Lasting Lessons of SARS, The" (Rosen), 147
Latin America, 2–3, 36–37, 76
Laveran, Charles Louis Alphonse, 78, 178*n*13
Lazear, Jesse, 17
Leeuwenhoek, Antoni Van, 139
*Legionella pneumophila* bacterium, 48
Legionnaires' disease, 48–49, 171*n*41
*Le Monde,* 193*n*6
lepers as scapegoats for plague, 9, 25, 161*n*13, 164*n*33–34
lepromatous leprosy, 74, 75
leprosy, 73–76, 177*n*10
leprosy bacillus, discovery of, 176*n*7
Leprovac vaccine, 177*n*10
leuco-agglutination, 183*n*3
Lewin, Roger Amos, 128–129
Liberia, 28, 45
Liesegang's rings, 127–128
linkage disequilibrium, 183*n*4
listeria, 150, 175*n*59
Lokta, Alfred J., 188*n*12
Lokta–Volterra model, 129, 188*n*12
London, Jack, x, xi
Lorenz, Edward, 123, 126–127
Louis IX of France, King, 6
Lovelock, James, xii, 123–125, 186*n*2
Lyme disease, 49–51, 171*n*42, 171*n*44
lymphocytes, 103–104, 110, 114, 141, 179*n*21, 190*n*28

M
Machupo virus *(Arenaviridae),* ix, 46
macrophages, 75, 102, 110
Madagascar, malaria in, 77, 82
mad cow disease, x, 51–53, 60, 149–150, 157, 171*n*45
mad cow disease prions, 51–53, 60
*Magic Mountain* (Mann), 70
major histocompatibility complex. *See* MHC
malaria
  anti-malarial measures, 82–83
  biology of, 78–81
  and cinhona tree, 77, 178*n*14–15
  coevolution of, 133
  development of protozoon plasmodium, 79–80
  and hemoglobin, 80, 178*n*17
  and lymphomas, 190*n*28
  mortality rate, ix–x
  return of, 76–78, 89–90, 96–97
  water, demography, and, 62
  See also *Anopheles* mosquito
mammals
  development of antibiotic resistance, 87–88, 180*n*32

domestic, 58
rats, 6, 10
*See also* marine mammals; mice; primates
Mandelbrot, Benoit, 127
Manila as origin of Reston Ebola, 43
man is the epidemic, 148, 149–158
Mann, Thomas, 70
Marburg fever, 44
Marburg virus *(Filoviridae),* ix, 44, 158
Margulis, Lynn, 126
marine mammals
dolphins, 129, 130–131, 188*n*14, 189*n*15, 189*n*16
fatal infections among, 131–132
Gaia or chaos, 129–131
sea lions, 129–130
seals, 129–131
theory and reality, 132–136
Marseilles, France, 161*n*15
Mars (planet), 186*n*2
mathematic modeling of epidemics, 134
Mbeki, Thabo, 152
McClintock, Barbara, 176*n*4
McCormick, Joseph, 59–60
MDR (multi-drug-resistant) tuberculosis, 71, 73
measles, 3, 107, 184*n*6–7
measles virus, human, 130–131, 188*n*13
medicine. *See* pharmaceuticals
Mediterranean dolphins, 130–131
Mendelian genetics, 105
meningitis, 50, 61, 70, 107, 134
meningococcus bacterium, 134
menopause, 141–142
Merck, 90
MetaGaia and chaos model, 136
methane gas, 186*n*2
methicillin-resistant *Staphylococcus aureus* (MRSA), 85–86
Mexico and *Conquista,* 2–3, 23–24, 108–109, 121
MHC (major histocompatibility complex)
and antigens, 110–111, 115
coevolution, 119
discovery of, 183*n*3
gene therapy for, 146
and graft rejection, 103
and HLA alleles, 107
mate selection of mice and, 185*n*12
monomorphism of, 110, 115
targeting for gene therapy, 146
and T lymphocytes, 103–104
mice
*Apodemus agrarius,* 57–58
*Calomys callosus,* ix, 46, 47, 58
diseases associated with, 58
white-footed, 50
Michel, Albin, 177*n*9
microbes
and autoimmune diseases, 138–139
belief in eradication of, vii–viii
and cancer, 136–138
evolution of, 67–68
gene modification capability, 20–21
mutations to, 20, 39, 67–68, 133, 169*n*29
pre-epidemic period of amplification, 20
resemblance between human tissues and, 186*n*16
resistance to antibiotics, viii
resistance to antimicrobial molecules, 176*n*3
tools for recognizing, 146
*See also specific microbes*
microproteins, prions as, x, 52
migration of plague-resistant rats, 10
Miller, Anne, 83

Miller, Judith, 154
*Mirage of Health, The* (Dubois), 41, 169*n*35
*Moeurs intimes du passé. Les fléaux de l'humanité* (Michel), 177*n*9
Moles, Abraham, 135
monkeypox, 5, 28, 160*n*6
Montague, Lady Mary Wortley, 4
Montaigne, 161*n*12
Mora, Giangiamo, 164*n*35
*Morbidity and Mortality Weekly Report* (CDC), 30
morbillivirus, 130–131, 188*n*14
Morin, Edgar, 187*n*5, 189*n*27
Morocco, polio in, 113
mortality rates
  for AIDS, 33–34, 67, 150–151
  of Amerindians during Spanish *Conquista*, 2–3, 23–24, 108–109, 121
  for cholera, 14, 162n21
  developed world versus developing world, 95
  for infectious diseases, 143
  for mad cow disease, 51, 172*n*46
  for malaria, ix–x
  of malaria, 82
  for measles, 184*n*6–7
  for plague, 8–9, 10, 23, 164*n*30
  for pneumococcal pneumonia, 84
  of sea lions, dolphins, and seals, 129–131
  for typhoid and yellow fever epidemic, 22–23
Moscovici, Serge, 92
mosquitos
  *Aedes*, 57, 175*n*57
  *Aedes aegypti*, 17, 18, 59, 163*n*26
  *Aedes africanus* and *A. simpsoni*, 18
  *Aedes albopictus*, 60
  *Aedes camptorhyncus*, 55
  apes and, 18, 56, 112–113
  and avian malaria, 119–120
  coevolution of, 133
  and DDT, 89–90
  and dengue fever, 175*n*57
  and Eastern equine encephalitis, 60
  ideal conditions for reproduction, 62, 82, 145, 179*n*23
  and Japanese encephalitis, 58
  pesticide-resistant, 79–80, 82, 89–90
  and rheumatoid fever, 55
  and Rift Valley fever, 57
  and Ross River virus, 55
  and yellow fever, 17–18, 18
  See also *Anopheles* mosquito
Moulin, Anne-Marie, 143
*Mount Dragon* (Preston), 29
Mourant, A. E., 116
movies, 26, 29, 165*n*36
Mozambique, 28, 90
MRSA (methicillin-resistant *Staphylococcus aureus*), 85–86
multi-drug-resistant (MDR) tuberculosis, 71, 73
multiresistant bacteria, viii
*Mycobacterium leprae* bacillus, 73–74, 75, 177*n*10
*Mycobacterium leprae* (Hansen's bacillus), 73, 75
*Mycobacterium tuberculosos* bacillus, 70, 71
*Myobacterium tuberculosis* (tuberculosis bacillus), 175*n*1
"myth of the virgin," 151
myxomatosis, 22, 116, 163*n*29
myxoma virus, 8, 22, 117, 163*n*29, 186*n*15

N

Nairobi, 37, 167*n*20
Nakicenovic, Nebojsa, 61
*Nana* (Zola), 1, 159n1

NASA, 145
National Institute of Allergy and Infectious Diseases (NIAID), 67
*Natural History of Quinines* (Weddell), 178*n*15
natural immunity, 103
natural selection theory of evolution, 142, 185*n*11
*Nature,* 105
*Neisseria meningitis* microbe, 61
*Neophoca hookeri* (Hooker sea lions), 129–130
neutralist theory of evolution, 185*n*11
*New Alliance, The* (Prigogine), 128
new diseases, 29–68, 96–97
  about, 53–54, 68
  Ebola, 29, 41–44, 169*n*36, 170*n*38
  hantavirus, 46–48, 54–55, 170*n*39–40
  Legionnaires' disease, 48–49, 171*n*41
  mad cow disease (BSE), x, 51–53, 60, 149–150, 157, 171*n*45
  Marburg fever, ix, 44, 158
  *See also* AIDS; HIV
New Mexico, 46–48, 54–55, 170*n*39
news media
  on AIDS, 34
  on Ebola, 44
  lack of pertinent information, 44, 170*n*38
  sensationalism preference, ix–x
New York, NY, Spanish flu pandemic, 15
New Zealand, 34, 129–130
NIAID (National Institute of Allergy and Infectious Diseases), 67
Nigeria, Lassa fever in, 45
Normalized Difference Vegetation Index (NVDI), 145
nosocomial infections, 84–85, 180*n*29
NVDI (Normalized Difference Vegetation Index), 145

O
Oberammergau, Bavaria (Germany), 25–26
OCP (Onchocerciasis Control Program), 181*n*37
Oklahoma, pneumonia in, 84
*Onchocerca volvulus* parasite, 90
onchocerciasis, 90–91
Onchocerciasis Control Program (OCP), 181*n*37
opportunistic infections, 30, 35, 39–40, 165*n*1
"Origin of AIDS, The" (Curtis), 151–152
origin of life, 187–188*n*9
Oropouche virus, 31, 62, 165–166*n*4
*Outbreak* (film), 29
outbreaks, 61–64, 75
  by microbes alone, 67–68
  *See also specific outbreaks*
overpopulation and disease, 8–9, 20, 42, 69, 72

P
*Pan troglodytes troglodytes* (chimpanzee), 31–32
Papua New Guinea, 82, 86, 173*n*49
Paraguay, hantavirus in, 48
parasitoses
  and AIDS, 39–40, 169*n*31
  carriers of, 134–135
  from deforestation, ix
  long-lasting interactions with, 134–135
  "new" infectious agents discovered, 1977-1991, 174*n*51
  *See also* malaria
Paris, France, 2, 162n21, 164*n*34

Patarroyo, Manuel, 82
pathocenosis, 56
pathogen diffusion
  blood transfusions, 66
  of Creutzfeld-Jakob disease, 51–53, 60, 173*n*49–50
  drug use and, 33, 34, 35, 63–64, 67, 166*n*11
  Ebola, 42
  of HIV/AIDS, 30, 32–34, 35, 36, 40
  of leprosy, 74–75
  ships' ballasts, 14, 21, 60, 175*n*56
  transportation of merchandise, 21
  from travel and migration, 21, 58–62, 63, 78, 90
  via blood transfusions, 66
  *See also* social disorders
pathogen diffusion in, 32, 166*n*12
pathogens responsible for infectious diseases, 175*n*59
PCBs (polychlorinated biphenyls), 189*n*15
PDV-1 (phocine distemper virus-1), 130–131
Pelletier, Pierre Joseph, 77, 178*n*15
penicillin, 84, 180*n*26
peptides strategy, 192*n*57
persecution of Jews and lepers, 9, 25, 161*n*13, 164*n*33–34
Peryilhe, Bernard, 137
pesticide-resistant mosquitos, 79–80, 82, 89–90
phagocytes, 102–103, 175*n*1
pharmaceuticals
  for AIDS treatment, 33–35, 37, 89, 167*n*14, 167*n*19
  antitubercular, 70–71
  antivirals, 88–89
  for Lassa virus, 45
  for malaria, 77, 82
  for onchocerciasis, 90–91
  for pneumonia, 84
Philippines, 14, 82, 162*n*22
*Phoca sibirica* (Lake Baikal seals), 130, 131
phocine distemper virus-1 (PDV-1), 130–131
phylogenetic analysis of influenza virus A, 16–17, 163*n*25
Piazza, Guglielmo, 164*n*35
Pineo, Lily, 45
plague
  AIDS immunity as result of, 38, 168*n*27
  and bacteriological warfare, 153
  demographic and economic collapse from, 10–11
  epidemics of, 7
  eschatological perspective, 8
  human susceptibility to, 8–9
  immunity to, 10
  persecution of Jews and lepers, 9, 25, 161*n*13, 164*n*33–34
  symptoms of, 102
  writers and films on, 6–7, 26, 160*n*7–10, 165*n*36
plague bacillus *(Yersinia pestis)*
  about, 6
  antibiotics resistant form of, 12
  modifications of, 20–21
  mutations to, 20, 68
  origin of, 7–8
  weakening of, 10
plankton, 14, 162*n*20
plasmids, 86
*Plasmodium*
  avian malaria, 119–120
  drug-resistant, 76, 77
  and Duffy blood group, 80, 116
  malaria-related species, 76
  protozoon, development of, 79–80
  and sickle-cell anemia, 80
  species infecting humans, 76

pneumococcal pneumonia, 84, 86, 180*n*28
pneumococcus bacterium, 83, 84, 180*n*24, 180*n*26–27
*Pneumocystis carinii,* 30, 182–183*n*1
pneumonia
AIDS-related, 30
dolphins with, 189*n*16
in elderly persons, 140
and penicillin, 83–84
pneumococcal pneumonia, 84, 86, 180*n*28
pneumococcus bacterium, 83, 84, 180*n*24, 180*n*26–27
similarity to Legionnaires' disease, 48
WHO statistics on, 84
Poincaré, Henri, 127, 128, 187*n*6
poliomyelitis, 112, 184*n*9
political consequences of plague, 11
political crises and epidemics, 21
polychlorinated biphenyls (PCBs), 189*n*15
polymorphism, 80, 119–120, 178*n*17
Pontiac fever, 48
poverty. *See* social disorders
pox viruses, 5
predator-prey model, 129
predisposition. *See* biological predisposition; genetic predisposition
pregnant women with AIDS, 37, 39, 167*n*22, 192*n*50
Preston, Douglas, 29
Preston, Richard, 29, 41
Prigogine, Ilya, 127, 128, 188*n*11
Prigogine dissipative system, 187–188*n*9
primates
and AIDS, 31, 151–152
and Creutzfeld-Jakob disease, 52
and deforestation, 56
and Ebola, 43–44
and leprosy, 74
and Marburg fever, 44
and SIV, 31–32, 132, 133
sooty mangabeys, 132–133
and yellow fever, 18
principle of undecidability, 128
prions (protineaceous infectious particle)
about, x
and blood transfusions, 66
illegal sale of meat with, 53, 150
nature of, 51–52
Prusiner prions, 52, 173*n*49
transmission of, 51–53, 60, 66
prison conditions and tuberculosis, 72
prostitutes and pathogen diffusion
in Africa, 32, 37, 40, 59–60, 151, 166*n*10, 167*n*20
in Asia, 33, 34, 35, 36, 63, 151
in Haiti, 33
protineaceous infectious particle. *See* prions
protozoon plasmodium, development of, 79–80
Prusiner, Stanley B., 52, 173*n*50
Prusiner prions, 173*n*49
*Pseudomonas* bacterium, 157
psychological consequences of plague, 11
*Ptychodiscus brevis* alga, 131–132
purpuric fever, 68

Q
quarantine, 43, 58–59, 164*n*32

R
*R. rattus* or *R. norvegicus* (rats), 6, 10
rabbits and myxomatosis, 22, 163*n*29
Rajneesh sect's bioterrorism, 156
rationality, absence of, 24–25
rats *(R. rattus* or *R. norvegicus),* 6, 10
Reed, Walter, 17, 163*n*26

refugees and spread of disease, 25, 61–62, 66–67, 179*n*23, 180*n*28
Reichholf, Josef, 117, 120
Reid, Ann, 5
religious institutions
  Catholicism, 11, 26, 42
  Ebola and, 42, 169*n*37
  effect of plague, 11, 26
  and leprosy, 74
  and Passion of Christ presentation in Bavaria, 25–26
  *See also* spirituality and culture
RENA VI network, 192*n*50
resistance
  about, 92–97
  to antimicrobial molecules, 176*n*3
  cost of, 90–91
  to leprosy, 74
  to parasites, 89–90
  of *Plasmodium falciparum,* 77
resistant bacteria
  about, viii, 71
  diarrhea from, 180*n*31
  Hansen's bacillus, 75
  *Myobacterium leprae,* 75
  peptides strategy against, 192*n*57
  plague bacilli, 12
  pneumococcal pneumonia, 84, 86
  *Shigella* diarrhea, 87
  *Staphylococcus,* 85–86
resistant forms of malaria, 82, 179*n*23
resistant humans, 4, 163*n*24
resistant mosquitos, ix–x, 89–90
resistant parasites, 76
resistant rats, 10
resistant viruses, 35–36, 89
Reston Ebola, 43–44
rheumatism, 84
rheumatoid fever, 55
rhinovirus, 182–183*n*1
ricin, 158
*Rickettsia tsusugamuchi,* 56
Rifkin, Jeremy, 193*n*6
Rift Valley fever, 57–58, 68, 174*n*53, 182*n*39
ring vaccination system, 194*n*9
*River, The* (Hooper), 151
rivet hypothesis of deforestation, 57
RNA, 16–17, 187*n*9
Rogers, Alan, 142
Rosen, Fred, 147, 192*n*59
Ross, Sir Ronald, 178*n*13
Ross River virus, 55
Rous, Peyton, 136–137
Ruffié, J., 109, 114, 140
Russia
  bacteriological warfare plans, 154–155, 194*n*8
  diphtheria, 21–22
Rwanda, 33, 61, 166*n*12

S
safe sex, 33, 34, 37, 39, 167*n*20
Salmon, Jean, 101–102
salmonella bioterrorism, 156
*Salmonella enteriditis* bacterium, 56, 65, 150
salmonella epidemic, 65
*Salmonella newport* bacterium, 88
Salt Lake City, UT, 84
Santa Fe Institute (SFI), 187*n*8
Santo Domingo, 24
sarin nerve gas bioterrorism, 156
SARS (Severe Acute Respiratory Syndrome), 15, 147, 192*n*53
satellites, 192*n*52
*Scarlet Plague, The* (London), x, xi
*Schistosoma japonicum* parasite, 106
*Schistosoma mansoni* parasite, 106
schistosomiasis, 106
*Science,* 133, 158
sea lions, 129–130
seals, 129–131
self-organization of complex phenomena, 189*n*20

self-organized criticality, 128
"sensitive dependency on initial condition," 127
septicemia, 180*n*25
Severe Acute Respiratory Syndrome (SARS), 15, 147, 192*n*53
sexually-related diseases
about, 63–64
safe sex, 33, 34, 37, 39, 167*n*20
syphilis, 22–23, 105, 109, 182*n*1, 182*n*38, 186*n*14
*See also* AIDS; HIV
sexual tourism, 175*n*54
SFI (Santa Fe Institute), 187*n*8
Shaw, Charlotte, 45
*Shigella* diarrhea, 87
sickle-cell anemia, 80, 178*n*16, 178*n*18
Sierra Leone, 15, 45
simian immunodeficiency virus (SIV), 31–32, 132, 133. *See also* primates
Sinclair, Upton, 150, 193*n*2
Sin Nombre virus (SNV) or hantavirus pulmonary syndrome, 46–48, 54–55, 170*n*39–40
Sir Lanka, malaria in, 77
SIV (simian immunodeficiency virus), 31–32, 132, 133. *See also* primates
skin and immunity, 102–104
skin graft compatibility, 80–81, 103–104, 179*n*21, 183*n*3
Skosana, Ben, 151
sleeping sickness, 117
slow-virus hypothesis for mad cow disease, 173*n*49
smallpox
AIDS immunity as result of, 38, 168*n*27
as bioterrorist weapon, 157
epidemics of, 2–4
eradication of disease, 4–5
human fear of, 1–2, 159*n*1
ring vaccination system, 194*n*9
smallpox virus, 2, 5, 159*n*2, 159*n*4, 194*n*8
*Smallpox 2002* (BBC), 157
Smallpox zero day, 4, 5
Snow, John, 161*n*18
SNV (hantavirus pulmonary syndrome or Sin Nombre virus), 46–48, 54–55, 170*n*39–40
social aspects of epidemics
about, 21–22
AIDS, 29
and BSE, 53
and demographic repercussions, 23–24
*See also* anarchical situations; biosociocultural aspects of epidemics; hygiene; immunity of populations; spirituality and culture
social disorders
and AIDS, 35
and cholera, 12–13, 14, 162*n*21
distribution of medicine issues, 90–91
and Ebola, 42, 169*n*37
and hantavirus, 47
and leprosy, 74
and malaria, 78, 81–82
overpopulation, 8–9, 20, 42, 69, 72
and polio, 184*n*9–10
refugees from, 25, 61–62, 66–67, 179*n*23, 180*n*28
social inequalities as infection cofactor, 34, 35, 67
socioeconomic instability, 62, 78
and susceptibility to disease, 21–22
*See also* hygiene; wars
socioeconomic instability, 62, 78
Solomon Islands, malaria in, 82

Somalia, 2, 28
sooty mangabeys, 132–133
South Africa, 36, 86, 152
South America
cholera pandemic, 14, 161*n*19
Eastern equine encephalitis, 60, 175*n*55
hantavirus, 48
yellow fever, 17
*See also specific South American countries*
Spanish colonization of Latin America, 2–3, 23–24, 108–109, 121
Spanish flu pandemic
about, 5, 15–16, 23–25, 162–163*n*
lack of notoriety, viii
species interaction in an ecosystem, 188*n*12
spirituality and culture
about, 25–27, 96–97
Daisyworld, 124–125
and origin of smallpox, 24
and plague, 24, 164*n*31
priests treating the sick, 26
retribution against Jews and lepers for plague, 9, 25, 161*n*13, 164*n*33–34
*See also* fear of epidemics
spontaneous disappearance of infectious diseases, 91–92
*Staphylococcus* bacterium, 84–85, 180*n*29
*Stegomyia* mosquito, 175*n*57
Stewart, William H., 146
stochastic theory of evolution, 185*n*11
*Streptococcus* bacterium, 83, 180*n*24, 180*n*26–27
streptomycin, 176*n*2
Sudan, malaria in, 82
Sudan Ebola, 41, 42
Surinam, typhoid and yellow fever epidemic, 22–23
susceptibility to cholera, 12
susceptibility to HIV, 37
sweating sickness epidemic, 91–92, 106
swine and influenza virus, 16, 163*n*25
syphilis, 22–23, 105, 109, 182*n*1, 182*n*38, 186*n*14

T
Taiwan, 142
Tanzania
cholera, 28, 182*n*39
HIV and AIDS, 32, 166*n*12
Taubenberger, Jeffrey, 5
Tchak, Sami, 151
technology and "progress," 64–67
Tempest, Bruce, 47
Tenochtitlán and *Conquista*, 2–3, 23–24, 108–109, 121
tetanus vaccinations, 107
Thailand
child prostitutes in, 33, 63
HIV and AIDS, 33–34, 35, 167*n*14, 179*n*23
malaria, 77, 82
thalassemia, 178*n*16
theory of natural selection, 185*n*11
theory of the inheritance of acquired characteristics, 185*n*11
therapy costs, 34
third horseman of the Apocalypse, 6. *See also* plague
three-body problem, 187*n*6
ticks *(Ixodes dammini, I. capillaris, I. pacificus, and I. ricinus)*, 49–50, 53, 171*n*44
*Time* (magazine), 37, 76, 169*n*38
*Times of Feast, Times of Famine* (Ladurie), 22
T lymphocytes, 103–104, 110, 114, 141, 179*n*21

TNF (tumor necrosis factor), 103
Tomkins, Clara, 173*n*50
*Topley and Wilson,* 56–57, 189
toxic shock syndrome, 85, 103, 180*n*29
transmissible prion-related diseases, 52–53
transposons, 176*n*4
travel and epidemics, 21, 58–62, 63, 78, 90
tri-therapy treatment for AIDS, 35–36, 37, 167*n*14
Troup, Jeannette, 45
*Trypanosoma gambiense, T. rhodesiense,* and *T. brucei,* 117
trypanosomiasis, 117
tubercular leprosy, 74
tubercular meningitis, 70
tuberculosis bacillus *(Myobacterium tuberculosis),* 175*n*1
tuberculosis (White Death), vii–viii, 69–73, 176*n*5
tumor necrosis factor (TNF), 103
Turkmenistan, diphtheria in, 21–22
typhus and typhoid, 22, 56, 121, 182*n*1, 186*n*14

U
Uganda
  and HIV, 36, 166*n*12
  jaw tumors of children, 137
  Marburg fever in primates from, 44
Ukraine, diphtheria in, 21–22
UNAIDS Compendium on Discrimination, Stigmatization, and Denial, 152
United Kingdom and BSE, 51–52
United Nations Development Program, 169*n*30
United Nations on world population, 142
United States
  biological resistance to HIV, 38
  cholera pandemics, 13
  Eastern equine encephalitis, 60, 175*n*55
  FoodNet program, 144–145
  HIV and AIDS, 34, 67
  Lyme disease, 49–50, 53, 171*n*44
  pneumonia, 84
  polio, 113
  small pox epidemics, 24
  Spanish flu pandemic, 15
  yellow fever epidemics, 18
  *See also specific states*
Uruguay, hantavirus in, 48
Utah
  hantavirus, 46–48, 54–55, 170*n*39
  pneumonia, 84

V
vaccination
  of elderly people, 139–140
  as hope for the future, viii, 184*n*8
  humans' survival of epidemics as, 22, 163*n*27
  and immunity of populations, 133–134
  for influenza, 17
  inoculation compared to, 4
  Leprovac vaccine, 177*n*10
  ring vaccination system, 194*n*9
  smallpox, as bioterrorism preparation, 157
Valen, Leigh Van, 132
vancomycin-resistant *Staphylococcus aureus* (VRSA), 85–86
Vanuatu, malaria in, 82
Venezuelan hemorrhagic fever, 174*n*51
Venice, Italy, plague statistics from, 10

*Vibrio cholerae. See* cholera bacterium
Vietnam, 179*n*22–23
Viking (Mars lander) problem, 186*n*2
viral infections, transmission method and seriousness of, 184*n*7
Virgil, 193*n*4
viruses
and cancers in humans, 137–138
from deforestation, ix
in human feces, 189*n*17
"new" infectious agents discovered, 1977-1991, 174*n*51
WHO on respiratory diseases caused by, 180*n*28
*See also specific viruses*
*virus qui détruisent les hommes, Ces* (Chastel), 5
vitamin B12 and cyanocobalamin, 180*n*32
Volterra, Vittorio, 188*n*12
Volterra model, 129, 188*n*12
VRSA (vancomycin-resistant *Staphylococcus aureus*), 85–86

W
wars
and AIDS pandemic in Africa, 33, 166*n*12
bacteriological warfare in WWI, 157
cholera pandemics associated with, 13
*Conquista,* 2–3, 23–24, 108–109, 121
destruction to balance of immunity of populations, 121
*See also* social disorders
water
cholera and, 13–14, 22, 161*n*18
demography and, 62
*Legionella pneumophila* bacterium and, 48–49
Watson–Crick pairing, 187*n*9
Web site information on epidemics, 144, 182*n*39, 191*n*46
Weddell, H.A., 178*n*15
White Death (tuberculosis), vii–viii, 69–73, 176*n*5
WHO. *See* World Health Organization
WHO (Global Malaria Eradication Program), 77
Wine, Laura, 45
women
and HIV/AIDS, 37, 39–40, 167*n*22, 169*n*30, 192*n*50
menopause, 141–142
pregnant, with AIDS, 37, 39, 167*n*22, 192*n*50
*See also* prostitutes and pathogen diffusion
World Health Organization (WHO)
on AIDS, 150–151
AIDS infection and mortality rates, 34, 167*n*13
on current incidence of malaria, 76
Global Malaria Eradication Program, 77
information on epidemics, 28, 182*n*39, 191*n*46
optimism in 1950s–1970s, vii
reported successes, 95
on respiratory infections in children, 84, 180*n*28
on smallpox eradication, 2
smallpox eradication program, 4–5
on tuberculosis, 73
on variola virus destruction, 160*n*5
Web site information on epidemics, 144, 182*n*39, 191*n*46
writings
on chaos theory, 188*n*11

on Ebola, 29
on leprosy, 177*n*9
on plague, 6–7, 160*n*7–10
on self-organization of complex phenomena, 189*n*20
on TB hospitals, 70
on unsanitary conditions, 150, 193*n*2

X
xenophobia. *See* fear of epidemics
*Xenopsylla cheopsis* (fleas), 6–7

Y
Yambuku, Zaire, 42, 169*n*37
yellow fever, 17–18, 22–23, 28, 59, 62, 163*n*26
*Yersinia pestis. See* plague bacillus

Z
Zaire Ebola, 41, 42
Zhabotinski's reaction, 127, 128
Zimbabwe, 39, 59, 166–167*n*12
Zola, 1, 159*n*1
Zur Hausen, Harald, 137, 138

# OTHER DANA PRESS BOOKS AND PERIODICALS

www.dana.org/books/press

## BOOKS FOR GENERAL READERS

## IMMUNOLOGY

FATAL SEQUENCE: The Killer Within

*Kevin J. Tracey, M.D.*

An easily understood account of the spiral of sepsis, a sometimes fatal crisis that most often affects patients fighting off nonfatal illnesses or injury. Tracey puts the scientific and medical story of sepsis in the context of his battle to save a burned baby, a sensitive telling of cutting-edge science.

Cloth 225 pp. 1-932594-06-X • $23.95
Paper 225 pp. 1-932594-09-4 • $12.95

## BRAIN and MIND

THE DANA GUIDE TO BRAIN HEALTH: A Practical Family Reference from Medical Experts (with CD-ROM)

*Floyd E. Bloom, M.D., M. Flint Beal, M.D., and David J. Kupfer, M.D., Editors*
*Foreword by William Safire*

The only complete, authoritative family-friendly guide to the brain's development, health, and disorders. *The Dana Guide to Brain Health* offers ready reference to our latest understanding of brain diseases as well as information to help you participate in your family's care. 16 full-color pages and more than 200 black-and-white illustrations.

Paper (with CD-ROM) 744 pp. 1-932594-10-8 • $25.00

THE CREATING BRAIN: *The Neuroscience of Genius*

*Nancy C. Andreasen, M.D., Ph.D.*

A noted psychiatrist and bestselling author explores how the brain achieves creative breakthroughs, including questions such as how creative people are different and the difference between genius and intelligence. She also describes how to develop our creative capacity. 33 illustrations/photos.

Cloth 197 pp. 1-932594-07-8 • $23.95

THE ETHICAL BRAIN

*Michael S. Gazzaniga, Ph.D.*

Explores how the lessons of neuroscience help resolve today's ethical dilemmas, ranging from when life begins to free will and criminal responsibility. The author, a pioneer in cognitive neuroscience, is a member of the President's Council on Bioethics.

Cloth 201 pp. 1-932594-01-9 • $25.00

A GOOD START IN LIFE: Understanding Your Child's Brain and Behavior from Birth to Age 6

*Norbert Herschkowitz, M.D., and Elinore Chapman Herschkowitz*

The authors show how brain development shapes a child's personality and behavior, discussing appropriate rule-setting, the child's moral sense, temperament, language, playing, aggression, impulse control, and empathy. 13 illustrations.

Cloth 283 pp. 0-309-07639-0 • $22.95
Paper (Updated with new material) 312 pp. 0-9723830-5-0 • $13.95

BACK FROM THE BRINK: How Crises Spur Doctors to New Discoveries about the Brain

*Edward J. Sylvester*

In two academic medical centers, Columbia's New York Presbyterian and Johns Hopkins Medical Institutions, a new breed of doctor, the neurointensivist, saves patients with life-threatening brain injuries. 16 illustrations/photos.

Cloth 296 pp. 0-9723830-4-2 • $25.00

THE BARD ON THE BRAIN: Understanding the Mind Through the Art of Shakespeare and the Science of Brain Imaging

*Paul Matthews, M.D., and Jeffrey McQuain, Ph.D.*
*Foreword by Diane Ackerman*

Explores the beauty and mystery of the human mind and the workings of the brain, following the path the Bard pointed out in 35 of the most famous speeches from his plays. 100 illustrations.

Cloth 248 pp. 0-9723830-2-6 • $35.00

STRIKING BACK AT STROKE: A Doctor-Patient Journal

*Cleo Hutton and Louis R. Caplan, M.D.*

A personal account with medical guidance from a leading neurologist for anyone enduring the changes that a stroke can bring to a life, a family, and a sense of self. 15 illustrations.

Cloth 240 pp. 0-9723830-1-8 • $27.00

UNDERSTANDING DEPRESSION: What We Know and What You Can Do About It

*J. Raymond DePaulo Jr., M.D., and Leslie Alan Horvitz.*
*Foreword by Kay Redfield Jamison, Ph.D.*

What depression is, who gets it and why, what happens in the brain, troubles that come with the illness, and the treatments that work.

Cloth 304 pp. 0-471-39552-8 • $24.95
Paper 296 pp. 0-471-43030-7 • $14.95

KEEP YOUR BRAIN YOUNG: The Complete Guide to Physical and Emotional Health and Longevity

*Guy McKhann, M.D., and Marilyn Albert, Ph.D.*

Every aspect of aging and the brain: changes in memory, nutrition, mood, sleep, and sex, as well as the later problems in alcohol use, vision, hearing, movement, and balance.

Cloth 304 pp. 0-471-40792-5 • $24.95
Paper 304 pp. 0-471-43028-5 • $15.95

THE END OF STRESS AS WE KNOW IT

*Bruce McEwen, Ph.D., with Elizabeth Norton Lasley*
*Foreword by Robert Sapolsky*

How brain and body work under stress and how it is possible to avoid its debilitating effects.

Cloth 239 pp. 0-309-07640-4 • $27.95
Paper 262 pp. 0-309-09121-7 • $19.95

IN SEARCH OF THE LOST CORD: Solving the Mystery of Spinal Cord Regeneration

*Luba Vikhanski*

The story of the scientists and science involved in the international scientific race to find ways to repair the damaged spinal cord and restore movement. 21 photos; 12 illustrations.

Cloth 269 pp. 0-309-07437-1 • $27.95

THE SECRET LIFE OF THE BRAIN

*Richard Restak, M.D.*
*Foreword by David Grubin*

Companion book to the PBS series of the same name, exploring recent discoveries about the brain from infancy through old age.

Cloth 201 pp. 0-309-07435-5 • $35.00

THE LONGEVITY STRATEGY: How to Live to 100 Using the Brain-Body Connection

*David Mahoney and Richard Restak, M.D.*
*Foreword by William Safire*

Advice on the brain and aging well.

Cloth 250 pp. 0-471-24867-3 • $22.95
Paper 272 pp. 0-471-32794-8 • $14.95

STATES OF MIND: New Discoveries about How Our Brains Make Us Who We Are

*Roberta Conlan, Editor*

Adapted from the Dana/Smithsonian Associates lecture series by eight of the country's top brain scientists, including the 2000 Nobel laureate in medicine, Eric Kandel.

Cloth 214 pp. 0-471-29963-4 • $24.95
Paper 224 pp. 0-471-39973-6 • $18.95

## THE DANA FOUNDATION SERIES ON NEUROETHICS

HARD SCIENCE, HARD CHOICES: Facts, Ethics, and Policies Guiding Brain Science Today

*Sandra J. Ackerman, Editor*

Top scholars and scientists discuss new and complex medical and social ethics brought about by advances in neuroscience. Based on an invitational meeting co-sponsored by the Library of Congress, the National Institutes of Health, the Columbia University Center for Bioethics, and the Dana Foundation.

Paper 152 pp. 1-932594-02-7 • $12.95

NEUROSCIENCE AND THE LAW: Brain, Mind, and the Scales of Justice

*Brent Garland, Editor. With commissioned papers by Michael S. Gazzaniga, Ph.D., and Megan S. Steven; Laurence R. Tancredi, M.D., J.D.; Henry T. Greely, J.D.; and Stephen J. Morse, J.D., Ph.D.*

How discoveries in neuroscience influence criminal and civil justice, based on an invitational meeting of 26 top neuroscientists, legal scholars, attorneys, and state and federal judges convened by the Dana Foundation and the American Association for the Advancement of Science.

Paper 226 pp. 1-932594-04-3 • $8.95

BEYOND THERAPY: Biotechnology and the Pursuit of Happiness. A Report of the President's Council on Bioethics

*Special Foreword by Leon R. Kass, M.D., Chairman.*
*Introduction by William Safire*

Can biotechnology satisfy human desires for better children, superior performance, ageless bodies, and happy souls? This report says these possibilities present us with profound ethical challenges and choices. Includes dissenting commentary by scientist members of the Council.

Paper 376 pp. 1-932594-05-1 • $10.95

NEUROETHICS: Mapping the Field. Conference Proceedings.

*Steven J. Marcus, Editor*

Proceedings of the landmark 2002 conference organized by Stanford University and the University of California, San Francisco, at which more than 150 neuroscientists, bioethicists, psychiatrists and psychologists, philosophers, and professors of law and public policy debated the ethical implications of neuroscience research findings. 50 illustrations.

Paper 367 pp. 0-9723830-0-X • $10.95

## ARTS EDUCATION

A WELL-TEMPERED MIND: Using Music to Help Children Listen and Learn

*Peter Perret and Janet Fox*
*Foreword by Maya Angelou*

Five musicians enter elementary school classrooms, helping children learn about music and contributing to both higher enthusiasm and improved academic performance. This charming story gives us a taste of things to come in one of the newest areas of brain research: the effect of music on the brain. 12 illustrations.

Cloth 231 pp. 1-932594-03-5 • $22.95
Paper 231 pp. 1-932594-08-6 • $12.00

## FREE EDUCATIONAL BOOKS

(Information about ordering and downloadable PDFs are available at www.dana.org.)

PARTNERING ARTS EDUCATION: A Working Model from ArtsConnection

This publication describes how classroom teachers and artists learned to form partnerships as they built successful residencies in schools. *Partnering Arts Education* provides insight and concrete steps in the ArtsConnection model. 55 pp.

ACTS OF ACHIEVEMENT: The Role of Performing Arts Centers in Education.

Profiles of more than 60 programs, plus eight extended case studies, from urban and rural communities across the United States, illustrating different approaches to performing arts education programs in school settings. Black-and-white photos throughout. 164 pp.

PLANNING AN ARTS-CENTERED SCHOOL: A Handbook

A practical guide for those interested in creating, maintaining, or upgrading arts-centered schools. Includes curriculum and development, governance, funding, assessment, and community participation. Black-and-white photos throughout. 164 pp.

THE DANA SOURCEBOOK OF BRAIN SCIENCE: Resources for Teachers and Students, Fourth Edition

A basic introduction to brain science, its history, current understanding of the brain, new developments, and future directions. 16 color photos; 29 black-and-white photos; 26 black-and- white illustrations.160 pp.

THE DANA SOURCEBOOK OF IMMUNOLOGY: Resources for Secondary and Post-Secondary Teachers and Students

An introduction to how the immune system protects us, what happens when it breaks down, the diseases that threaten it, and the unique relationship between the immune system and the brain. 5 color photos; 36 black-and-white photos; 11 black-and-white illustrations. 116 pp. ISSN: 1558-6758

## PERIODICALS

Dana Press also offers several periodicals dealing with arts education, immunology, and brain science. These periodicals are available free to subscribers by mail. Please visit www.dana.org.